Alexandra Hofmänner

Science & Technology Studies Elsewhere

A Postcolonial Programme

Schwabe Verlag

The prepress of this publication was funded by the Swiss National Science Foundation. Published with the support of the Freiwillige Akademische Gesellschaft. A research grant from the Max Geldner Foundation is gratefully acknowledged.

MIX
Papier aus verantwor-
tungsvollen Quellen
FSC® C083411

Bibliographic information published by the Deutsche Nationalbibliothek
The Deutsche Nationalbibliothek lists this publication in the Deutsche Nationalbibliografie; detailed bibliographic data are available on the Internet at http://dnb.dnb.de.

Cover illustration: Photography reproduced with kind permission of the artist Izel Lutz; Karoo, South Africa.
Copy-editing: Katherine Bird, Berlin
Cover design: icona basel gmbh, Basel
Graphic design: icona basel gmbh, Basel
Typesetting: 3w+p, Rimpar
Print: CPI books GmbH, Leck
Printed in Germany
ISBN Print 978-3-7574-0034-7
ISBN eBook (PDF) 978-3-7574-0039-2
DOI 10.24894/978-3-7574-0039-2
The ebook has identical page numbers to the print edition (first printing) and supports full-text search. Furthermore, the table of contents is linked to the headings.

rights@schwabeverlag.de
www.schwabeverlag.de

To Mias

Contents

Preface

In the many years it has taken to complete this book, I have received help from too many people to mention in a preface.

I would like to express my special thanks and appreciation to Helga Nowotny, Aant Elzinga, and Yehuda Elkana, whose pioneering work has in many respects paved the way for this study. They have encouraged me to investigate further, rather than surrender to, the formidable weight of history on today's worldwide knowledge and power constellations. I am also greatly indebted to the work of Edward Said, Michel-Rolph Trouillot, Gurminder Bhambra, Shalini Randeria and Stuart Hall. I am privileged to have worked with many extraordinary academics in South Africa and Switzerland who continue their work undaunted by obstacles and challenges and would like to thank them for their commitment which has been an inspiration for this book.

An earlier version of this book was submitted for Habilitation at the University of Basel. I would like to thank Sabine Maasen, Rayvon Fouché, and Elisio Macamo for their comments on the original manuscript and Agnes Hess, Edwin Constable, Klaus Neumann-Braun, Eveline Peterer and Moritz Maurer for their support in this process. A version of Chapter Three was published by Taylor & Francis in *Social Epistemology, A Journal of Knowledge, Culture and Policy* (2016, 30/2, pp. 186–212). I am indebted to my editors at the Schwabe Verlag for their assistance in the publication process.

I wish to thank Bruce Lewenstein and the Department of Science & Technology Studies at Cornell University for kindly hosting me and affording me the time and space to review the manuscript.

Finally, I would like to take this opportunity to express gratitude to my family and friends; this book would not have been possible without their support and encouragement.

Any remaining errors are entirely mine.

1. The Global March for Science, and STS

We live in an age that calls into question things that were previously tak-en for granted. The freedom of science is increasingly put under pressure internationally – and in Europe. And although science in the 21ˢᵗ century is as important as never before in view of pressing global challenges such as climate change, preservation of biodiversity, or combating infectious disease, there is a sense that fact-based knowledge is questioned more and more. This is unacceptable in the name of science as well as in the name of civil societies. We bear the responsibility. The March for Science is an opportunity to make this visible.
Prof. Martin Stratmann, President of the Max Planck Society
March for Science, Munich, April 2017.[1]

On 22 April 2017, scientists took to the streets in Silicon Valley, a symbol for leading-edge science and technology. They joined protestors in more than five hundred cities across the world on this day to participate in the March for Science. Demonstrating scientists were photographed carrying banners with slogans that read 'Science is not Fiction', 'Trust Scientific Facts Not Alternative Facts', 'Expert is not a dirty word', 'Science is Real'. These catch-words were repeated in media coverage on the event. The March for Science made it to newspaper headlines with titles like 'Scientists to take to the streets in global march for truth'.[2] The venerated scientific journals *Science* and *Nature*, and many reputable scientific institutions and associations such as

---

1   https://www.mpg.de/11221750/max-planck-praesident-spricht-beim-muenchner-march-for-science (author's translation).
2   This title is taken from an article by Mark Lynas as published in the British news-paper *The Guardian* on 17 April 2017.

the American Association for the Advancement of Science (AAAS), the German Science Council, or the Swiss Academy of Arts and Sciences, publicly endorsed the March for Science.

The idea to hold a protest March for Science is reported to have evolved from discussions on social network platforms over an initial plan to demonstrate in Washington D.C. (Ley et al., 2018). Within a very short time, the idea spread via Reddit, Facebook, and Twitter, and so-called satellite marches were organized in other cities around the world, linked through an internet website on the March for Science.

Why did the scientists take to the streets? The organizers of the March for Science expressed their concern that science and evidence-based policy were threatened by a loss of public confidence and scientific authority. The majority of the protestors defended the position that science is a public good, they believed that science informs responsible government policies and they supported government investments in science (Ross et al., 2018: 228). Proponents of the march advocated for an increase in activities in science outreach and communication, science education, and scientific literacy as means to defend science and evidence-based policy (Ross et al., 2018). The march was to mark the beginnings of a global movement to defend the important role of science in matters of health, security, economy and government.[3]

The March for Science was openly declared a political movement. The political activism was also criticized by other parts the scientific community. Concerns were raised about potential losses to the status of scientists as impartial experts, and about possible accusations that scientists are self-interested opportunists advocating for a cause to get more funding and prestige (Ross et al., 2018: 228).

In interviews, historians of science pronounced the march an unprecedented event in the history of science.[4] Indeed, no past event appeared to match the impressive number of scientists from across the spectrum of disciplines that took to the streets simultaneously in so many cities. Accordingly, the march was interpreted by the organizers, the media, and the

---

3   https://www.marchforscience.com/.

4   Chris Mooney. 2017. 'Historians say the March for Science is "pretty unprecedented"' Washington Post, 22 April 2017.

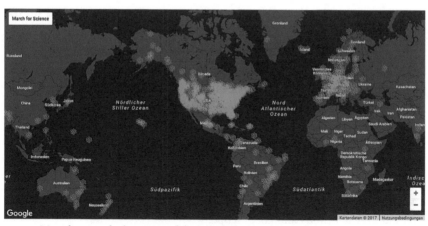

Figure 1: Map showing the locations of the March for Science in 2017.

endorsing scientific institutions as exposing a crisis in the relationship between science and society. The crisis had to do with the decline of public trust in science and the rise of 'post-truth' politics. In the preceding years, scientists had been active in citizen science projects, science cafés, participatory technology assessments, or public communication initiatives in an effort to reconfigure the relationship. Still, scientists took to the streets in April 2017 to demonstrate against the momentous changes that were transforming knowledge, power and truth in society.

Two further Marches for Science have taken place since then, but with far fewer participants. The sense of anti-science attitudes among the public and policymakers, and public scepticism about scientific evidence and expertise, however, has anything but diminished. On the contrary, these attitudes have increasingly come to be linked to recent political developments in Western liberal democracies, polarized politics and the rise of populism. There is widespread consensus that the relationship between science and society is in peril.

Against this background, the March for Science in 2017 raises a number of interesting questions. Why did these scientists take to the streets rather than use conventional channels of science advocacy? And why did the event take place in 2017? What lies behind this apparent loss of public confidence in science? Why is scientific knowledge seen to be losing authority?

The relationship between science and society has been studied by the academic field of Science & Technology Studies (STS) since the late 1960s and early 1970s. STS draws from a variety of disciplines, theories and approaches to study the advancement of science and technology and their interaction with the many spheres of social life. In the words of the most recent *Handbook of Science and Technology Studies*, STS 'investigates the institutions, practices, meanings, and outcomes of science and technology and their multiple entanglements with the worlds people inhabit, their lives, and their values' (Felt et al., 2018: 1). Over the past fifty years, STS has gained a rightful place in academia and has assembled analytical and theoretical tools for studying science, technology and society. The March for Science falls squarely within its analytical scope.

Given the scale of participation in the March for Science in 2017, and the nature of the issues raised, STS is arguably well suited to turn to for guidance, analysis and interpretation of our questions on the March for Science. However, the prominent STS journals issued no articles to analyse the event. Neither has the March emerged as a key topic at the conferences of STS's two main professional associations.[5] At the same time, however, the issues raised at this event continue to occupy social and political life in Western liberal democracies. So why has STS not seized upon the March for Science as a moment to showcase its analytical competences and potential contribution to contemporary affairs? Why did the March for Science not make it onto the research agenda of STS?

The answer lies in a dilemma that has come to haunt the field of STS. As an academic field, STS has developed and employed analytical tools and frameworks that have contributed to the demystification of scientific and technological expertise by disclosing their historical, political, social and other contingencies. With the advent of 'post-truth' politics and the rising public critique of experts, STS has been blamed 'for the onset of the post-truth era' (Sismondo, 2017: 587). The dilemma for STS scholars has to do with their simultaneous roles as scientific experts and as critics of scientific expertise.

---

[5]   The Society for the Social Study of Science (4S) and the European Association for the Study of Science and Technology (EASST) are considered here as the two main professional associations for STS.

As experts on matters concerning science, technology and society, they belong to the community of academics who defend the exclusivity and validity of scientific knowledge. As experts on matters concerning science, technology and society, however, they also know that the standards and norms that attribute scientific exclusivity and validity are inherently contingent and change over time. The March for Science confronts STS with this dilemma. The crisis in the relationship between science and society, which prompted scientists to take to the streets and protest against the decline of public trust in science and 'post-truth' politics, has also plunged STS into a crisis of legitimacy (Sismondo, 2017; Collins et al., 2018; Fuller, 2016; Jasanoff & Simmet, 2017).

The premise of this book is that the crisis of legitimacy affords an opportunity for STS to refurbish its analytical toolkit and conceptual frameworks. The book suggests that one of the reasons why STS has not produced analyses and interpretations of the March for Science is that this field is not conceptually equipped to carry out this task. It claims that postcolonial perspectives on science, technology and society offer new paths for interpreting the meaning of the March for Science. This can be achieved by expanding the current debate in STS on the crisis in the relationship between science and society beyond the conclusion of a universal decline of public trust in science and the rise of 'post-truth' politics.

The call for postcolonial perspectives in STS might easily be averted by pointing to the recent international expansion of STS and the concomitant multiplication of situated knowledges across the globe, as professed in the latest *Handbook of Science and Technology Studies* (2017). However, the alleged international expansion in geographic terms appears not to translate into the 'epistemic landscape' of STS; the *Handbook* itself declares that approximately ninety per cent of the *Handbook's* authors are affiliated to institutions in the United States and Europe, with over two thirds in the United States and the United Kingdom (Felt et al., 2017). Therefore, references to the geographic expansion of STS tells us little about its effect on the analytical tools and conceptual frameworks in the field of STS.

The argument for postcolonial perspectives in STS is not new but dates back to the early 1990s. However, postcolonial science studies, or postcolonial technoscience, has remained an addendum to mainstream STS, one of numerous approaches that might be pursued in the study of science, technology and society, particularly in postcolonial geographical settings. Rather than a mere addendum, this book argues that the postcolonial perspective can contribute to tackling the momentous changes that are currently transforming knowledge, power and truth in society by revising and expanding the conceptual and analytical toolkit of STS. To put the postcolonial approach to work, the book proposes a Programme in Science Studies Elsewhere. The programme is developed on the basis of an examination of two seminal books that influenced the formation and course of STS in the mid-1980s: an in-depth empirical study of the seminal book by Thomas P. Hughes, *Networks of Power. Electrification in Western Society, 1880–1930* (1983), and an analysis of the influential book by Steven Shapin and Simon Schaffer, *Leviathan and the Air Pump. Hobbes, Boyle and the Experimental Life* (1985).

The study begins with a brief report on STS's new wave of expansion to settings beyond Western Europe and North America (Chapter Two). The chapter also introduces the paradox of the Collective Blind Spot of STS, raises the question of its implications for the recent expansion of the field of STS and describes the particular postcolonial approach to be employed in this study. The next chapter applies this approach to Hughes' study on the history of electrification in Chicago, Berlin and London, by adding the fourth example of Johannesburg (Chapter Three). Chapter Four presents a postcolonial examination of Shapin and Schaffer's study of experimental practice in early modern science. The findings of the postcolonial analyses are then put to work in the Programme in Science Studies Elsewhere which is formulated in Chapter Five. Chapter Six discusses the implications of the postcolonial analysis for STS, and the last chapter returns to the March for Science to draw conclusions from the study for the field of STS (Chapter Seven).

2. Science and Technology Studies Goes Global

The topic of this book may be introduced by way of a mental picture. Imagine the historian of technology, Thomas P. Hughes, standing on the fiftieth floor of the Carlton Centre skyscraper in Johannesburg (also called the 'Top of Africa') enjoying the panoramic view of the city of gold. In its heyday, the Carlton Centre was the tallest building in the Southern hemisphere and a symbol of wealth and status. Hughes has just published his celebrated 1983 book, *Networks of Power: Electrification in Western Society, 1880–1930*. He had studied the history of electrification in Chicago, Berlin and London, and presented a model of technological change that would leave an imprint on subsequent academic studies of technology. What would Thomas P. Hughes see? What questions would he ask about the history of the electrification of Johannesburg, inspired by this bird's eye view?

Looking south beyond the city centre, the landscape reveals the residue of more than a century of mining along the gold reef outcrop. The remains of mine dumps effectively display the origins of the city as a gold mining camp. Gold was discovered in the wider metropolitan area of Johannesburg (also known as the Witwatersrand) in 1886. Within barely a decade, the Witwatersrand was producing one-fifth of the world's gold output. By 1925, the output from these mines had risen to more than half of the world's gold production. Within this time frame, even before the First World War, one of the world's largest electric power schemes was built to provide electric power for the gold mines – before regional power systems were established in Berlin or London.

From his lookout point, Hughes might note that the electricity transmission lines only covered selected residential areas of Johannesburg. In the mid-1980s, only roughly one third of households in South Africa would have had access to electricity. The gigantic electricity scheme that had been built in Johannesburg at the beginning of the twentieth century had led to patterns of electricity distribution that overlooked the majority of the population. With his model of technological change in mind, how would Hughes imagine the history of electrification in Johannesburg? How would it relate to his work on the history of electrification in Chicago, Berlin and London?

Hughes' seminal book is attributed to the academic field of Science and Technology Studies (STS). This field emerged in the late 1960s and early 1970s, and over the past fifty years has achieved a legitimate interdisciplinary position among the contested territories of academic disciplines: journals and international and national academic associations have been established, handbooks have been published, university degree courses are offered, and core literature and themes have been recognized. Hughes' study of technology contributed to shaping this emerging field in the mid-1980s. His model introduced new concepts to the study of technology that continue to influence the field of STS today.

The STS researcher who sets out to study the history of electrification in South Africa is likely to turn to Hughes' systems model of technological change for methodical guidance. He or she is also likely to consult other theories of technological change that developed in STS subsequent to Hughes such as, for example, Actor Network Theory (ANT). However, there is no getting around Hughes' history of electrification in Western society as reference to study the history of electrification in South Africa. Indeed, to omit Hughes's study might compromise this scholar's qualification as a member of the STS community.

The field of STS has secured a reputable academic position for studying scientific and technological change. It has become an academic resource that is consulted to address contemporary problems that concern science, technology and society. Today, science and technology are still regarded as the most promising endeavours for finding solutions to the numerous problems of humankind. At the same time, the rapid pace at which science and technology penetrate our daily lives gives rise to continual scepticism and criticisms. Over the past few years, there have been increasing calls to pay attention to the new forces appearing on the global map of scientific and technological excellence, which is said to be undergoing a transformation.[6] Western European countries have developed new national and international science and technology policies to respond to these changing international conditions.[7] Likewise, the March for Science in 2017, as an historically unprecedented event in the history of science, supports the premise that new forces are currently reshaping the relationship between science, technology and society.

More than ever, the field of STS is being challenged to help us to understand and respond to these rapidly changing circumstances. Therefore, it is important to ask how well this field is equipped to deal with these changing global circumstances. STS scholars might respond by pointing out the field's international expansion. New STS communities and initiatives have been observed to be emerging in various geographic regions such as, for example, East Asia. STS scholars might also submit that new academic journals have been founded, new scientific associations have been formed, and new university courses are being taught in places beyond Europe and North America.

STS's growth in numerical terms across geographical territories, however, tells us little about how well STS is equipped intellectually to cope with the global challenges described above. Despite its relative success as an academic field, STS continues to be plagued by a number of problems. As an example, there are fierce internal battles for epistemic sovereignty and disciplinary authority over the study of science and technology, despite STS's self-declared interdisciplinary disposition. STS's role in the decline of public trust in science and 'post-truth' politics has also given rise to discussions in

---

6   This changing political economy of science and technology is typically described by quantitative indicators for a range of issues including e.g. the emergence of new players (such as Brazil, Russia, India, China, South Africa and South Korea); new patterns of international research investment (particularly in emerging economies); new information and communication technologies; access to knowledge; new forms of competition (i.e. from brain drain to brain circulation); new international regulations; and an increasing number of internationally co-authored scientific publications and mobility of researchers (e.g. NAS, 2011; EC, 2012; OECD, 2011; Royal Society, 2010; UNESCO, 2010).

7   See, for example, OECD 2005, 2010a, 2010b, 2011 and 2012; European Commission, 2008 and 2012; and U.S. National Academy of Sciences (NAS), 2011.

the STS community. Furthermore, the silence of STS with regard to the historically unprecedented March for Science, as a manifestation of a crisis between science and society, suggests that the crisis has also left its mark on STS as a field of scientific expertise.

Against this background, this book aims to contribute to the development of STS by looking more closely at what happens when this field, as an intellectual programme, goes global. It starts from the assumption that this project requires attention to be given to one of STS's predicaments. This predicament has been noted by various STS scholars but has yet to receive much attention in the academic debate. Let us call this predicament 'the paradox' of STS, as it has to do with a fundamental contradiction that is implicitly maintained in the field. The paradox is likely to haunt any STS scholar standing on the fiftieth floor of the 'Top of Africa' skyscraper in Johannesburg, contemplating science and technology from this vantage point. He or she will search the intellectual repertoire of STS for concepts, models, approaches, theories and methods to make sense of science and technology in this scenery. For the case of electrification, the STS scholar is likely to end up using the resources of the intellectual traditions of Actor Network Theory (ANT), Large Technological Systems (LTS), or the Social Construction of Technology (SCOT). The paradox enters as soon as this conceptual repertoire is mobilized.

2.1 The Paradox: STS's Collective Blind Spot

The academic field of STS has successfully achieved a legitimate standing among other fields of knowledge. To secure this academic identity, STS scholars have navigated the obligatory rites of passage that are marked out by the gatekeepers of knowledge. Fields of knowledge attain academic standing by presenting a history of intellectual achievements. This intellectual history is an important proof of the validity of knowledge produced because it has to stand the test of intellectual independence, untouched by politically, economically or socially vested interests. Indeed, it is precisely this detached status that makes an academic field trustworthy.

In the case of STS, however, the academic stipulation for intellectual independence creates a critical dilemma. The field of STS itself has criticized the image of the virtuous progression of knowledge as a cumulative intellectual enterprise. It has successfully demonstrated the intricate ways in which knowledge is intertwined with, and co-constituted by, its specific contexts. In 1997, the parting president of the European Association for the Study of Science and Technology (EASST), Aant Elzinga described this *paradox*:

> STS-people are puzzling. They tend to become rather narrow-minded in constructing their own past, positioning themselves programmatically in mostly cognitive terms, while ignoring their own broader historical and socio-political contextuality. I guess this is part of the (very much "modern") game of having to legitimate oneself in purely professional terms in academe in the fight for respectability and funding (Elzinga, 1997).

By this act, STS scholars typically 'tell their own origins story so as to emphasize the internal development of their history, and to neglect any version of externalism [...] or any social constructionist account', while they simultaneously 'take for granted the social context of science'. Elzinga then notes the tendency to regard Thomas Kuhn as the 'founding father' of STS and as 'single-handedly opening the doors to the possibility of a fully social account of science'. He questions this account and insists 'that attention is paid both to theories and their historical location' (Elzinga, 1997).[8]

---

8   Elzinga refers to the previous work of Loren Graham and Hilary Rose to make his point: '[...] Loren Graham remarked how scholars in our field have repealed internalism but when they write their own history in the long term perspective they sometimes point to J.D. Bernal and even Joseph Needham, and so they play up the event of the History of Science Congress in London 1931 as very significant. Then they go on to trace an intellectual genealogy from the ideas in Boris Hessen's benchmark paper to Bernal 1939, Needham, Levy Hyman, and others to the externalism-internalism debate in the 1960s (with Merton coming in as a target alongside historians like Butterfield) – leaving out all the activism in and around the key figures from the thirties. Thereafter Kuhn's The Structure is usually introduced as a new benchmark. In this reconstruction of the past something vital is rendered invisible. It is the existence and importance of the Science and its Social Relations movement of the 1930s and the many science activists in various countries who took part as committed and reflective intellectuals and scientists in a broader anti-facist and anti-racist movement of that time. Their concern, shared with Merton's, was "keeping science straight". Hilary Rose for her part has noted in the same way how the "social

A few years later, Guggenheim and Nowotny (2003) introduced the notion of the *Collective Blind Spot* of STS to capture this paradoxical condition.

> Science Studies' standard genealogy, its academic self-portrait, stands in stark contrast to its own intellectual programme in which it has compellingly criticized the representation of knowledge fields as a cumulative sequence of intellectual achievements and individual biographies. Indeed, such Whig historiography of science is diametrically opposed to STS's canonical argument of co-construction between research and subject (Guggenheim & Nowotny, 2003: 239).

Guggenheim and Nowonty point out that the Collective Blind Spot of STS might lead to 'fateful misconceptions, uncritically favouring the actions of heroic individuals and promoting a notion of progress as inherent in the field' (Guggenheim & Nowotny, 2003: 235–6).

This contradiction places the STS scholar in a paradoxical intellectual and professional position. Although this paradox has been noted repeatedly, it has not received much attention. However, it has not been swept under the carpet; on the contrary, the problem is acknowledged by mainstream STS scholars. Lynch's introduction to his edited four volumes on *Science and Technology Studies: Critical Concepts in the Social Sciences* ends with an elegy on STS's dilemma of progress.

> Despite the questioning of the very idea of scientific progress in post-Kuhnian STS, narratives of progress have been a conspicuous part of the field's development over the past several decades. There is no question that STS has become more visible and institutionally established both within, and to some extent beyond, academia. This progress – with the tendentious narratives that accompany and celebrate it – faces a central dilemma. The dilemma is that the standard polemics for promoting a field, gathering adherents and envisioning future success are endemic to the positive philosophy that STS cut its teeth against (Lynch, 2012: 49).

Incidentally, Lynch forgot about this dilemma and its consequences when he compiled the texts for the four-volume opus to demarcate the epistemic

---

turn" in our later history from the late 60s onward too has been reconstructed as an internalist cognitive genealogy that tends to fix upon Kuhn' (Elzinga, 1997).

territories of the field of STS. He also disremembered it in the volume's introduction which offers an (intellectual) genealogy of STS concepts and approaches. Thus, the acknowledgement of STS's paradox only pays intellectual lip service to the problem.

This book assumes that the paradox of STS's Collective Blind Spot has implications for the international travels of STS. It claims that postcolonial perspectives can shed new light on these relations. Before introducing the postcolonial approach that will be applied in this book, a short profile of STS is offered below.

2.2 What is STS?

STS is typically introduced as an interdisciplinary academic field with origins in Western Europe and North America in the late 1960s or early 1970s. The common assumption underlying the emergence of this field is that science and technology have to be studied as social activities. In the words of Sergio Sismondo, 'the field investigates how scientific knowledge and technological artefacts are constructed' (Sismondo, 2010: 11). This section will briefly sketch the field of STS by recounting the categories and standard narratives that are typically used to introduce and legitimize the field of STS within disciplinary academic structures and beyond.

Mainstream descriptions of STS have tended to accentuate an unfolding *intellectual* history: a roadmap of evident academic strides marked by scholarly achievements and a number of intellectual 'turns'.[9] As already stated by Elzinga above, the historical roots of STS are typically traced to Thomas S. Kuhn's (1962) book on the *Structure of Scientific Revolutions*.[10] Rip describes Kuhn as 'something of a godfather to the field' (Rip, 1999). His book is usually credited with having instigated this new field of research by integrating 'historical, sociological and philosophical analysis' (Rip, 1999), but an earlier

---

[9]   See Webster, 1991; Swidler & Arditi, 1994; Edge, 1995; Biagioli, 1999; Collins & Evans, 2002; Bucchi, 2004; Yearley, 2005; Bauchspies et al., 2006; Sismondo, 2008; Turner, 2008; or Lynch, 2012.

[10]   See Restivo, 1987; Rip, 1999; Sismondo, 2008, 2010; Turner, 2008; Martin et al., 2012; and Lynch, 2012.

intellectual heritage is often acknowledged: '[...] the intellectual imprint of S&T Studies, and its sense of breaking new ground, becomes recognizable at that time' (Rip, 1999).

Other accounts trace STS's intellectual heritage back beyond the 1960s,[11] promote an earlier group of founding fathers,[12] or simply refer to earlier scholars to describe the 'pre-history' of STS.[13] Although the names and dates may be a matter of dispute, the standard format for the founding myth of STS consists of references to a limited number of intellectual forefathers who have paved the way for the academic progression of STS.[14] On the shoulders

---

[11]   For example, Felt et al. in their chapter 'Eine kurze Geschichte der Wissenschafts-forschung' refer to Saint-Simon, Karl Marx, Friedrich Engels (historischer Materia-lismus), Max Weber's Gesellschaftstheorie, Max Scheler & Karl Mannheim (Wissens-soziologie), Alfred Lotka (Szientometrie), Boris Hessen, Ossowska/Ossowski, Popper/Polanyi, Bernal, Needham, and Merton (Felt et al., 1995).

[12]   David Edge's classic historical account of STS in the introductory chapter to the second *Handbook of Science & Technology Studies* in 1995 traces STS's origins back to the 'mood of the mid-1960s' (Edge, 1995: 6). Edge's history of STS starts out by recalling the intellectual influence of Derek de Solla Price and his seminal book Little Science, Big Science 1963, and then proceeds to refer to a line of scholars ranging from John D. Bernal, Michael Polanyi, and Thomas Kuhn to Robert K. Merton. Stephen Cutcliffe refers to C.P. Snow as 'Perhaps the most influential intellectual precursor of the STS movement' (Cutcliffe, 2000: 7). He considers Snow's metaphor of 'two cultures' to serve as an impor-tant reference point for the field of STS.

[13]   The *Handbook on Science and Technology Studies* (2008) exemplifies this historical approach to presenting STS. Its first chapter (by Sergio Sismondo) charts the intellectual emergence of the field from Kuhn's book. Its second chapter (by Steven Turner) is devo-ted to 'The Social Study of Science before Kuhn'. This chapter traces the heritage that led to the emergence of the field, since, 'The controversy over Thomas Kuhn's astonishingly successful Structure of Scientific Revolutions [...] created the conditions for producing the field that became "science studies"' (Turner, 2008: 33).

[14]   This approach to STS history has materialized in introductions to volumes that compile key texts for the field of STS. Mario Biagioli's edited *Science Studies Reader* (1999) is a case in point. His volume is offered for use 'as the core text in introductory courses in science studies at the graduate and undergraduate level' (Biagioli, 1999: XIV). His history of STS presents this field's intellectual achievement. Although he omits from this edited collection 'the work of such classic authors as Fleck, Kuhn, Feyerabend, Mer-ton, Canguilhem, Foucault, Barnes, Bloor, and Bachelard', he strongly encourages readers

of these intellectual founding fathers, the subsequent development of the field is recounted along a few key historical markers, including its inter-disciplinary approach; the development of research approaches of increasing sophistication (such as the Sociology of Scientific Knowledge (SSK), the Social Construction of Technology (SCOT), the Empirical Programme of Relativism (EPOR), or Actor Network Theory (ANT) etc.); the identification of classic essays; its case study focus; its institutional history; and its division into relativist and policy-oriented camps, etc.

At the most basic level, STS's history is characterized by its integration of a variety of disciplinary approaches and recurrent debates on their respective territories.[15] Sismondo's Introduction to STS describes STS as 'a dynamic interdisciplinary field [...] the result of the intersection of work by sociologists, historians, philosophers, anthropologists, and others studying the processes and outcomes of science, including medical science, and technology' (Sismondo, 2010: vii). Cutcliffe (2000) identifies changes in various 'traditional disciplinary academic fields' that prompted 'historians, sociologists, and philosophers of both science and technology increasingly [to move] away from internalist-oriented subdisciplines to progressively more externalist or 'contextual' interpretations'. (Cutcliffe, 2000: 7).

Another standard trajectory for recounting the history of STS has been the intellectual development of *traditions* or *subfields* of increasing sophistication (Lynch, 2012; Edge, 1995). In this perspective, the development of STS maps an increasing scope of inquiry and expertise, '[...] starting with

---

'to familiarize themselves with this literature as it is very relevant to understanding the intellectual genealogies of the essays presented here' (Biagioli, 1999: XVI).

15   For example, the *Advanced Introduction to Science Studies* by Hess (1997) notes that, 'Notwithstanding the growth of interdisciplinarity, the disciplinary divisions remain strong, and they underlie the organization of this book. Chapters 2, 3, and 4 are therefore organized as introductions to the philosophy of science, the sociology of science, and the sociology of scientific knowledge. These fields still constitute the major sources of specialist terminology and theorizing. The title of chapter 5, "Critical and Cultural Studies of Science and Technology" is suggestive of my view of where the field is moving. This chapter introduces concepts from a number of overlapping fields: anthropology, critical social theory, cultural studies, feminist studies, critical technology studies, and the cultural history of science' (Hess, 1997: 3).

scientific knowledge, and expanding to artefacts, methods, materials, observations, phenomena, classifications, institutions, interests, histories and cultures.' (Sismondo, 2008: 13). The notion of construction as a central feature of STS plays a decisive role in this point of view. Indeed, some scholars see the history of STS as a progressive history of 'extensions of constructivist approaches' (Sismondo, 2008: 17).

> The metaphor of constructivism, in its generic form, thus ties together much of STS: Kuhn's historiography of Science; the strong program's rejection of non-naturalist explanations; ethnographic interest in the stabilization of materials and knowledges; EPOR's insistence on the muteness of the objects of study; historical epistemology's exploration of even the most apparently basic concepts, methods, and ideals; SCOT's observation of the interpretive flexibility of even the most straightforward of technologies; ANT's mandate to distribute the agency of technoscience widely: and the co-productionist attention to simultaneous work on technical and social orders. [...] Yet the metaphor has enough substance to help distinguish STS from more general history of science and technology, for the rationalist project of philosophy of science, from the phenomenological tradition of philosophy of technology, and from the constraints of institutional sociology of science (Sismondo, 2008: 17).

One more strategy to map STS as an academic field has been seen in the form of collections of *classic essays* in STS (examples are Biagioli, 1999 and Lynch, 2012). Often, these essays follow the case study format, which also is sometimes referred to as a common empirical feature of STS: 'Case studies are the bread and butter of STS. Almost all insights in the field grow out of them, and researchers and students still turn to articles based on cases to learn central ideas and to puzzle thought problems. The empirical examples used in this book point to a number of canonical and useful studies' (Sismondo, 2010: viii). When, at times, the *institutional history* of STS is mentioned, it usually remains confined to a selected chronology of the dates of origin of the first STS programmes at universities in the US[16] or Europe. Standard histories of STS typically do not refer to the formative political,

---

16   'Other founding STS programs with interdisciplinary curricular orientations included those at Cornell University (1969), Pennsylvania State University (1968/69), Stanford University (1970/71), and Lehigh University (1972) [...]' (Cutcliffe, 1989: 422).

social or economic conditions of the early STS programmes, organizations and publications.[17]

An additional typical feature of STS that is mentioned in introductory texts is its division into two distinct camps or 'two subcultures – one activist, the other academic' (Waks, 1993: 339).[18] Fuller (1993) and Fuller and Collier (2004) further characterize STS by designating these camps as the high and low church tendencies.

> 'High Church' STS tends to be interested in the special epistemic status that science enjoys vis-à-vis other forms of knowledge. In coming to understand how science organizes itself internally and projects itself externally, STS began mimicking those very processes to acquire academic respectability and expert authority. In contrast, 'Low Church' STS focuses more on the problems that science has caused and solved in modern society. From the Low Church standpoint, STS was preoccupied with proliferating jargon, establishing self-contained citation networks, and solidifying a canon [...]. Both the High Church and Low Church sects of STS like to trace their origins to the 1960s. Whereas the High Church points to Thomas Kuhn's The Structure of Scientific Revolutions (1970) as the watershed STS text, the Low Church portrays STS as a response to the disturbing symbiosis that developed between scientific research and the military establishment during the Vietnam War' (Fuller 2000b) (Fuller and Collier, 2004 [1993], xii).

These two subcultures also point to an alternative perspective on the emergence, identity and purpose of STS, which sees the field as having a dual heritage that is intellectual and activist. Although some scholars do not count the policy-oriented stream as a valid part of STS, others acknowledge these two perspectives but afford the intellectual perspective epistemic sovereignty over the assimilated activist stances.[19] A third group of scholars view STS's

---

17   For example, Cutcliffe's mention of a 'five-million dollar grant from IBM' to establish the Harvard University Program on Technology and Society in 1964, almost stands out as a breach of taboo (Cutcliffe, 1989: 422 and Cutcliffe, 2000: 10).

18   Quoting Ilerbaig's contrast (Ilerbaig, J. 1992: the Two STS Sub-Cultures and the Sociological Revolution. Science, Technology and Society, Vol. 90, pp. 1–5).

19   'Parallel to this in the academy, "Science, Technology and Society" became, starting in the 1970s, the label for a diverse group united by progressive goals and an interest in science and technology as problematic social institutions. For researchers on Science, Technology and Society the project of understanding the social nature of science has gen-

activist origins as constitutive of the field. For example, Aant Elzinga refers to a co-evolution of intellectual and social dimensions that shaped STS's heritage when he contemplates the forces at play in the establishment of the European Association for the Study of Science and Technology (EASST) (Elzinga, 1997). He mentions various social movements in the 1960s and 1970s that involved 'criticism of science, its uses and its products on the one hand, but questioning also the very value of instrumental rationality', such as 'the anti-imperialist movement of those days, the radical student protest movements, the environmental movement, and particularly the women's movement'. He traces the roots of STS within a broader perspective of movements for the social responsibility of science beyond 1970 to the 1930s and projects such as setting up the Committee on Science and its Social Relations (CSSR) under the International Council of Scientific Unions (ICSU) (Elzinga, 1997). In 2003, Michael Guggenheim and Helga Nowotny again emphasized the importance of this dual heritage[20] and referred to the establishment of the social sciences in 19th-century Europe as an example of how STS's history reflects an already well-known 'coevolutionary development between social movements and their subsequent academic integration or appropriation' (Guggenheim & Nowotny, 2003: 8).[21]

---

erally been seen as continuous with the project of promoting a socially responsible science (ref to Ravetz, Spiegel Rösing Price, Cutcliffe). [...] This is the other "STS," which has played a major role in Science and Technology Studies, the former being both an antecedent of and now a part of the latter' (Sismondo, 2010: 10).

20   STS in this view grew at least partly out of the broad stream of other political and social movements of the late sixties and early seventies, as did feminism and the environmental movement (Guggenheim & Nowotny, 2003: 8).

21   Stephen Cutcliffe's introduction to STS also accentuates the activist roots of STS in his description of the origins of two early STS programmes at North American universities: 'One of the first [programs] was the Science, Technology, and Society program at Cornell University, which appeared in 1969 at least in part as a response to campus unrest and the need to develop "interdisciplinary courses at the undergraduate level on topics relevant to the world's problems."' (Cutcliffe, 2000: 10) and, 'Another important early program – the Science, Technology, and Society Program at Pennsylvania State University – emerged out of a "Two Cultures Dialog" begun in 1968/69. It solidified about 1971, under the influence of the Cornell program. For many years it served as the host institu-

Overall, however, current introductions to the field adhere to the intellectual historiography that sketches the progressive introduction of scientific concepts, approaches and theories (e.g., Webster, 1991; Edge, 1995; Biagioli, 1999; Collins & Evans, 2002; Bucchi, 2004; Yearley, 2005; Turner, 2008; Lynch, 2012). Mainstream accounts of STS tend to overlook in their history the influence of pioneering scholars who have made significant contributions to the institutionalization of STS within academia. Standard histories of STS have favoured the field's intellectual heritage over its activist or 'low church' camp heritage.[22] This 'taming of science studies by its academic context' (Martin, 1993: 255) feeds into the preservation of the Collective Blind Spot of STS. The insistence on an intellectual history also affords the field of STS a sheltered place among European and North American knowledge institutions and disciplines. But what happens when STS goes global?

2.3 Transporting the Collective Blind Spot of STS

This book is concerned with the implications of the Collective Blind Spot of STS for its development in geographical regions beyond Europe and North America, henceforth subsumed under the expression 'the North Atlantic' (Trouillot, 2002a).[23] As mentioned above, the genealogy of STS charts its intellectual emergence in the North Atlantic, from where it subsequently spread to other geographical regions.[24] Again, Michael Lynch's introduction

---

tion for the National Association for Science, Technology and Society' (Cutcliffe, 2000: 11).

[22]   Brian Martin noted this omission in an article on 'The Critique of Science Becomes Academic' in 1993: 'Lack of acknowledgement of radical or activist origins is symptomatic of the process by which academics use the critique of science for professional purposes, distancing it from working scientists and activists dealing with the impact of science. The process of academization has seen a move from a critique of science in society to a critique of scientific knowledge and finally to a critique of the knower' (Martin, 1993: 251).

[23]   'Science and Technology Studies (STS) is a dynamic interdisciplinary field, rapidly becoming established in North America and Europe' (Sismondo, 2010: vii).

[24]   This type of genealogy is described by Stephen Cutcliffe: 'To assist in carrying out that mission, numerous STS programs have come into being during the past three decades. While the specific number is not clearly known, and some have fallen along the way-

to the authoritative four volumes on *Science and Technology Studies: Critical Concepts in the Social Sciences* classically reproduces this picture: 'The geographical base of the field also broadened as STS programmes and associations formed in East Asia, India, South America and elsewhere' (Lynch, 2012: 5).

However, there are alternative perspectives on the genealogy of STS and its geographic expansion. For example, scholars from postcolonial science and technology studies have pointed out that scholarly work on science and technology has long been carried out in other geographical regions, but it has not found its way into the STS canonical literature (Harding, 1998; McNeil, 2005[25]). Others actually consider the early attempts to join forces to study science and technology beyond the geographical boundaries of Europe and North America as one of driving forces for the foundation of the field (Spiegel-Rösing & Price, 1977; Elzinga, 1997; Cutcliffe, 2000).

Lynch's claim that STS has 'attained a growing international profile in recent decades' (Lynch, 2012: 1), however, is not corroborated by the record, as evidenced by the most recent *Handbook of Science and Technology Studies* (2017). The great majority of the STS scholars who defined the epistemic terrain of the field of STS in the Handbook are associated to institutions in the United States and Europe. No authors from 'Australia or Africa' and only seven from Asia and South America have contributed to the chapters of the *Handbook* (Felt et al., 2017: 15, 16). This skewed geography of the epistemic

---

side, the number of full-fledged programs in the United States numbers nearly one hundred, with perhaps a similar number in Europe. Equally important are the hundreds of individual courses and groups of courses, which, while they cannot be considered programs in the fullest sense, certainly complement the more formally established programs. Similar program and course development has also take place in Japan, China, Canada, Australia and several Latin American nations' (Cutcliffe, 2000: 10).

25   'Nevertheless, it is important to register that too often science and technology studies have lacked or marginalised global perspectives. In origin and focus the field has been predominantly Eurocentric, influenced by the actors and concerns of Europe and its diasporic communities in North America and Australia. Nevertheless, there have been critical strains within STS which have highlighted the 'provincial' nature of such orientations' (McNeil, 2005: 106).

landscape of STS[26] raises a number of questions about STS's specific capacity to study contemporary global dynamics in science and technology.

The journal of *East Asian Science, Technology and Society* (EASTS), established in 2007, offers excellent examples for examining some of these questions. EASTS 'aims to bring together East Asian and Western scholars from the fields of science, technology, and society (STS)'.[27] The Journal 'serves as a gathering place to facilitate the growing efforts of STS networks from Northeast Asia, Southeast Asia, North America, and Europe to foster an internationally open and inclusive community'.[28] In 2008, historian of science Warwick Anderson described the establishment of EASTS as a signal for '[…] the emergence of novel sites for STS and the development of a broader community of scholars' that 'provides a guide to the travels of STS beyond Western Europe and North America' (Anderson, 2007: 249). At stake in this enterprise, according to Anderson, was nothing less than 'to re-chart the map of STS for the twenty-first century' (Anderson, 2007: 249).

This image of an emerging field of STS in East Asia provides a good opportunity to inquire into the effects of STS's Collective Blind Spot on the 'travels of STS' beyond the North Atlantic. Anderson's vision for a map of STS symbolizes Elzinga's paradox: the 'travels of STS' beyond the North Atlantic replicate the archetypal image of scientific fields as composed of disembodied concepts and ideas that can be transported across contexts and time. The legitimacy of this image has been criticized – and indeed literally overthrown – by STS itself in recent decades. Nevertheless, Anderson's vision of STS expanding on a global scale can be seen to represent mainstream views. For example, in 2010, the annual meeting of the Society for

---

26   For example, Ben Martin et al. have analysed the core literature of STS over the past fifty years to identify the twenty most influential contributions to the field by counting the average number of citations per year in the Web of Science. The national origin of twelve of these contributions are the United States, seven are from the UK, three from France (one author) and two from the Netherlands (Martin et al., 2012: 8). This analysis was conducted by means of a bibliometric study starting out with five handbooks that have been published in the field as reference works (Martin et al., 2012: 6).

27   EASTS is sponsored by the Ministry of Science and Technology, Taiwan.

28   The conference was jointly organized by 4S and the Japanese Society for Science and Technology Studies.

Social Studies of Science (4S) was held in Asia for the first time, and the conference was entitled 'STS in Global Contexts'. The description of the conference reads as follows: '4S members will have a chance to experience, interact with and understand the cultural diversity of Asia. Furthermore, holding 4S in Asia opens the door to questions relating to universalities and cultural differences in STS concepts. This meeting is an example for reconsidering STS in global contexts as well as strengthening STS network worldwide'.[29]

Some of the consequences of Elzinga's paradox of STS are apparent in a set of questions raised by the scholar Fa-ti Fan in the second issue of the new EASTS journal: 'Is East Asia a useful category for science and technology studies? Is East Asian STS simply the application of existing theories from the United States or Europe to East Asia? Is its aim simply to produce case studies modelled on Western scholarship? How can East Asian STS be fruitfully distinctive from what is being practiced in the West today?' (Fan, 2007: 243). In the same issue, the scholar Hideto Nakajima compares the history of STS in Japan with that in East Asia and concludes that 'Japanese STS is now a mixture of European and American STSs', whereas STS scholars in other regions in East Asia 'are under stronger influence, or almost sole dominance, by American STS' (Nakajima, 2007: 240).

Fa-ti Fan's thoughts are significant in the light of the issues raised earlier about STS's contribution to the study of global dynamics of science and technology, and to the March for Science in 2017 in particular. Indeed, they appear to present a critical stumbling block to practising STS outside of the North Atlantic. Writing about the joint conference of 4S and EASTS in 2008, Amit Prasad summarized these problems by formulating his version of STS's Collective Blind Spot.

> It was, however, striking how often presenters and interjectors [...] argued for the need of STS to travel to such regions without acknowledging that this may perhaps require a change in STS analytics. It seemed as though STSers, while consistently analysing the problematic of how science travels as 'immutable mobiles', have started to believe that STS could (or should) do the same (Prasad, 2008: 36).

---

29   http://www.4sonline.org/meeting/10 /accessed in August 2019.

2.4 A Postcolonial Lens: Science and Technology as North Atlantic Universals

We are left to surmise that Fa-ti Fan's questions cannot be reasonably answered without first coming to terms with STS's Collective Blind Spot. How can these questions be addressed? How can the paradox of the Collective Blind Spot of STS be tackled? Guggenheim and Nowotny (2003: 8) propose 'an assessment of STS in terms of its own standards and criteria and suggest writing a socio-cultural history of STS that takes into account questions such as 'What is the connection between the development and institutionalization of STS and broader economic and political developments? How did the specific cultural and geographic patterns of STS and its diversity in orientation and research style develop as they did?' (Guggenheim & Nowotny, 2003: 7).[30] The particular genealogy of STS as a North Atlantic field of knowledge, however, implies that a socio-cultural history as proposed by these authors (as important and urgent as it might be for other purposes) is compelled to remain trapped in the vicious circle of the Collective Blind Spot: examining STS in terms of its own standards and criteria will not eliminate the blind spot because STS is unable to overcome the co-construction of research and subject.

Anderson's portrayal of STS's travels provides a point of entry for this book to address STS's Collective Blind Spot. What kind of map does Anderson plot when he sees STS travel beyond Western Europe and North America to East Asia? His map plots STS in its standard intellectual genealogy, as an unfolding of an intellectual history with distinct national footprints that presents the work of North Atlantic scholars.[31] From its origins in Western

---

30   Additional questions raised by Guggenheim and Nowotny are: 'Why are there relatively many women in STS as compared to, say, economics, history, or even sociology and anthropology? What were the factors leading to the establishment of STS units in the universities of some countries but not in others? What were the relations between STS and science policy and what has changed, if anything, with the growing importance of EU research policies? Why did the political strand inside STS undergo a relative decline?' (Guggenheim & Nowotny, 2003: 7).

31   See Webster, 1991; Edge, 1995; Biagioli, 1999; Collins & Evans, 2002; Bucchi, 2004; Yearley, 2005; Turner, 2008; Lynch, 2012.

Europe and North America, the ideas and concepts of STS are seen to be relocated, translated and assembled in local contexts, such as 'Asia'.

This book focuses on STS's reciprocal co-construction of research and subject to investigate the implications of STS's Collective Blind Spot on the field's global disposition. However, Anderson's map of STS's travels beyond the North Atlantic does not simply chart the passage of an intellectual corpus of research, such as STS concepts, approaches or theories. It also quietly monitors the transference of STS's particular conceptions of its objects of study, science and technology. STS scholars project the universal validity of their own intellectual programmes on the global scale even though they bear particular historical imprints.

Michel-Rolph Trouillot has introduced the phrase *North Atlantic universals* to refer to words that project the experiences of the North Atlantic onto the global stage of history while simultaneously silencing their local heritages. *North Atlantic universals* designate a family of words such as development, progress, democracy and 'the West' (Trouillot, 2002a: 220). Trouillot regards these words as 'not merely descriptive or referential' because they do not describe the world but rather offer visions of the world. However, they are inherently ambiguous and can be seen as 'analytical fictions'.

> North Atlantic universals are particulars that have gained a degree of universality, chunks of human history that have become historical standards. They do not describe the world; they offer visions of the world. They appear to refer to things as they exist, but because they are rooted in a particular history, they evoke multiple layers of sensibilities, persuasions, cultural assumptions, and ideological choices tied to that localized history. They come to us loaded with aesthetic and stylistic sensibilities; religious and philosophical persuasions; cultural assumptions ranging from what it means to be a human being to the proper relationship between humans and the natural world; ideological choices ranging from the nature of the political to its possibilities of transformation. There is no unanimity within the North Atlantic itself on any of these issues, but there is a shared history of how these issues have been and should be debated, and these words carry that history. And yet, since they are projected as universals, they deny their localization, the sensibilities and the history from which they spring (Trouillot, 2003: 35).

Incidentally, *North Atlantic universals* display three additional properties: they are prescriptive, seductive, and difficult to conceptualize. They are prescriptive 'inasmuch as they always suggest a correct state of affairs – what is good, what is just, what is sublime or desirable' (Trouillot, 2003: 35). They are seductive, 'at times even irresistible', because they have the capacity 'to project clarity while remaining ambiguous' (Trouillot, 2002a: 221). This capacity, in turn, makes them problematic to conceptualize (Trouillot, 2002a).

North Atlantic universals may be studied by mapping two related geographies, or 'two complementary spaces': *the geography of imagination* and *the geography of management*. Whereas the geography of management creates places, the geography of imagination 'has to do with the relationship between place and space'. It necessarily requires a reading of alterity as 'a referent outside of itself—a pre- or non-modern in relation to which the modern takes its full meaning' (Trouillot, 2002a: 222). Trouillot offers the concept of '*Elsewhere*' to study this alterity. Elsewhere designates 'a space of and for the Other that can be, and often is, imaginary' (Trouillot, 2002a: 225). He contrasts 'the Here and the Elsewhere, which [premise] one another and [are] conceived as inseparable' (Trouillot, 2002a: 222).

Notably, Elsewhere designates a *space* rather than *place*, which encompasses both the (often imaginary) Other outside and the Other within. This kind of shift in perspective is effectively described in Marie Louise Pratt's analysis of travel writings about encounters with 'Bushmen' in Southern Africa in the 18th century. She claims that this kind of encounter not only involves a confrontation of Europeans with 'unfamiliar Others' but also involves a confrontation with 'unfamiliar selves' (Pratt, 1992: 140) because they are not explicitly anchored '[…] either in an observing self or in a particular encounter in which contact with the other takes place' (Pratt, 1992: 140). The travel writings are a mode of 'Othering', a normalizing discourse 'whose work is to codify difference […]' (Pratt, 1992: 139). Similarly, Stuart Hall has stated that although the Other appears to be 'banished to the edge of the conceptual world, constructed as absolute opposite, negation', it concurrently appears at the very centre (Hall, 1992: 221).

The example of Anderson's map of STS travels beyond Europe and North America suggests that STS's designated objects of study, science and technology, might be regarded as *North Atlantic universals* in Trouillot's sense. The term *North Atlantic universal* offers plenty of room for theoretical specification and empirical exploration, but Trouillot proposes two related geographies or lenses to study these kinds of words: the geography of management and the geography of imagination (Trouillot, 2002a).

This book assumes that STS's objects of study, science and technology, can be studied as *North Atlantic universals* through the two related lenses of the geography of management and the geography of imagination, and that this analysis will deliver new insights into STS's co-construction of research and subject. This study hypothesizes that by using Trouillot's lenses to examine case studies of key intellectual contributions to the field of STS, we can gain the empirical scope to develop new concepts to address STS's Collective Blind Spot. In their guise as *North Atlantic universals*, technology and science silence their particular histories and origins, and set standards of admission at the entry gates to the world stage of technology and science. This book is interested in three implications of these properties of STS's objects of study: their effects on STS's co-constructed body of research, concepts, methods and theories; the consequences for their global travels beyond the North Atlantic; and their connection to STS's Collective Blind Spot.

In other words, this book does not address STS' Collective Blind Spot by composing its socio-cultural history; rather, it sets out to explore STS's co-construction of research and subject further through Trouillot's postcolonial notion of the *North Atlantic universal*. The co-construction of STS's particular conceptions of science and technology and the field's intellectual products will be examined by means of two seminal books: *Networks of Power: Electrification in Western Society 1880–1930* (1983) by Thomas Hughes, and *Leviathan and the Air-Pump: Hobbes, Boyle, and the Experimental Life* by Shapin and Schaffer (2001 [1985]). The geographies of management and imagination will be employed to analyse both books, so they need to be specified in more detail. Trouillot's description of these notions is converted into a number of key questions to guide the inquiry (see Table 1).

It is important to qualify the nature and purpose of the postcolonial approach employed in this study, lest it be assigned to some radical anti-STS programme. Much STS work undertakes empirical research to test and further develop the canonical conceptual landscape of STS. This book undertakes a journey to critically consider this conceptual landscape. Its purpose is not to disprove the brilliant work of Hughes or of Shapin and Schaffer. On

Geography of management	What procedures and institutions of control were elaborated and implemented both at home and abroad?
	What is their relationship to the development of world capitalism, which reorganizes space for explicitly political or economic purposes?
Geography of imagination	What ideas and (individual and collective) identities inhabit the geography of imagination?
	Which two complementary spaces (not places) of the Here and the Elsewhere (and its Other) have been used continually to recreate the West?
Geographies of management and imagination	How are the geographies of management and imagination intertwined, and what is their relationship? In which ways are they inseparable, and do they premise each other?
	How do the geographies of management and imagination underpin historicism and the legitimacy of the West as the universal unmarked?
	How do the *North Atlantic universals* of technology and science set the terms of the debate and restrict the range of possible responses?

Table 1: Key questions for the empirical study of North Atlantic universals with the lenses of the geographies of management and imagination (based on Trouillot, 1991; 2002b).

the contrary, had they not produced such sophisticated intellectual work, a systematic study would not have been possible. Neither is this book intended to criticize STS. It seeks to contribute to current issues and the future development of this field, which (like any other academic field) has developed in a specific manner, although many other paths were possible.

STS would likely have looked different today without Hughes' *Networks of Power* or Shapin and Schaffer's *Leviathan and the Air-Pump*. They coined concepts and approaches that went on to mark some of the central standards and unique achievements that constitute the field. Some authors have left stronger imprints than others on the academic standards that have been used to manage participation in this field of study. This book maintains that it is the business of STS to critically consider, from time to time, the analytical tools that define these standards against new historical conditions and prospects. This is particularly important when STS 'goes global', as pointed out by Amit Prasad, '[...] we need to explore whether (and along with it also why) STS needs a retooling of its analytics when it shifts its focus to the trans-national or global arena' (Prasad, 2008: 36).

This book also does not aim to undermine the legitimacy of STS's empirical body of research, which has applied these concepts and theories.

Rather, it seeks to help reshape the conceptual landscape in order to constantly consider the best fit between STS's body of research, concepts, approaches and theories on the one hand and the problems STS researchers are trying to solve on the other hand. In this sense, the conceptual categories that compose STS's body of research will be treated as transient categories that shape historical thinking. The ultimate purpose is to position the field more strongly as an intellectual field of inquiry within the global social sciences and humanities. This book is not yet another exercise in deconstruction. Its critical approach is inspired by Yehuda Elkana's unrelenting question: 'Are we conceptually equipped to deal with the world?' Elkana was adamant that there is no strict separation between theory and practice, because conceptual frameworks are translated by carriers of knowledge of all kinds into daily reality. Conversely, reality is read through conceptual frameworks.

> What remains to us? To admit that we do not have, and never had, all embracing theories of anything and to look realistically for the best and broadest local theories that are practically useful, even if among them, for the time being or forever, contradictions prevail. For that we need new concepts and a basic reeducation of our thinking to a dialectical ability to live with contradictions in our theories; to live with contradictions in practice we are used to and this constitutes no difficulty, but to tolerate them in our theoretical frameworks is a radically new ballgame (Elkana, 2000: 18).

Following in Elkana's footsteps, which draw no strict dividing line between theory and practice, this book does not begin with a theoretical discussion of selected STS concepts set against Trouillot's notion of the *North Atlantic universal*. Instead, two influential STS studies on science and technology will be reconsidered by applying the lens of the *North Atlantic universal*. The objective is to revisit the concepts put forward in these STS studies some thirty years after their introduction and to suggest alternative analytical tools with which to study science and technology that cut across the conceptual landscape that codifies them as *North Atlantic universals*. Trouillot's spirit of investigation sets the tone for this project.

> But if the seduction of North Atlantic universals also has to do with their power to silence their own history, then we need to unearth those silences, the conceptual and theoretical missing links that make them so attractive (Trouillot, 2003: 36).

Let us briefly return to the introductory image of Thomas Hughes standing on the 'Top of Africa' skyscraper with a panoramic view of the city of gold, pondering the history of electrification in Johannesburg. This book assumes that Thomas Hughes would be wearing his particular scholarly spectacles with glasses that were fitted for his study of electrification in Western society – Chicago, Berlin and London. Through these lenses, technology would come into sight in its guise as a *North Atlantic universal*, and the history of electrification in Johannesburg would appear as a short, faulty chapter in the history of technology transfer to the peripheries of the West. Of interest to this book, however, is the Blind Spot that has settled on Hughes' spectacles, and its subsequent collective survival in the field of STS.

3. Technological Change and the "Savage Slot"

3.1 Thomas Hughes' *Electrification in Western Society*

This chapter offers a postcolonial reading of a book published in 1983 by Thomas Parker Hughes, an American historian of technology. Hughes' seminal book *Networks of Power: Electrification in Western Society 1880–1930* introduced a new approach to the study of technology and is generally acclaimed as one of the first interdisciplinary studies of technological change in the scholarly tradition of STS. The book formed part of the emerging new field of the sociology of technology, which in the mid-1980s started branching out into various subfields, such as the Social Construction of Technology (SCOT), Actor Network Theory (ANT) and Large Technological Systems (LTS). Hughes' analysis of electrification presented a number of terms and concepts that fostered these emerging academic debates and influenced the subsequent academic discourse on technology. For this reason, Hughes' study has secured an unchallenged status in standard STS genealogies and features as one of the foundational texts of the field of STS. As a pioneering interdisciplinary book, Hughes' study also contributed to securing STS-distinct academic terrain amid the contested territories of disciplinary fields. However, Hughes' *Networks of Power* has repercussions for the study of electrification and technological change beyond the field of STS. It has become a standard reference across a variety of academic disciplines, such as economics, sociology, history and philosophy. Leslie Hannah's review of the book in 1984 accurately predicted that 'the discussion of electric power systems will never be the same again' (Hannah, 1984: 382).

Hughes presented a model that aspired to capture the key agents, elements and dynamics of technological change. He is credited with having introduced a number of concepts that still circulate in empirical analysis in STS, such as large technological systems, the system builder, the reverse salient, critical problems, technological style and technological momentum. The object of investigation in Hughes' history of electrification is the electric power system. He claimed that the 'change in configuration of electric power systems 1880–1930' constituted the 'formative years of the history of electric supply systems' (Hughes, 1983: 1). Accordingly, his book pursued the question: 'How did the small lighting systems of the 1880s evolve into the regional power systems of the 1920s?' (Hughes, 1983: 2). One of the pioneering assumptions of Hughes' investigation into technological change was that the history of electric power systems extends 'beyond national borders' (Hughes, 1983: x). This assumption informed the empirical analysis upon which his model was based. He undertook a comparative analysis of electrification in three countries: Germany (Berlin), England (London) and America (Chicago). The objective of his book, therefore, was 'to explain the change in configuration of electric power systems across these three sites during the half-century between 1880 and 1930' (Hughes, 1983: 2).

Chicago, Berlin, London – Johannesburg

Hughes' history of electrification draws upon historical sources on Thomas Alva Edison's work and on the electrification in Chicago, Berlin and London.[32] Hughes' historical account emphasizes their differences with the intention of developing an overarching model of technological change that is valid beyond these specific empirical sites.[33] Indeed, his model proceeded to

---

[32] Hughes' explains his reasons for selecting Germany as follows: 'I found that the networks of evolving technologies often linked Germany with the United States because both were industrializing rapidly'. Hughes selected England because it 'often provided a contrast to events and trends observed in the other two countries' (Hughes, 1983: x). His reasons for restricting the study to three countries were 'Limitations of time, resources, and language prevented exploration of the sources pertaining to France, Italy, Sweden, the Benelux countries, Russia, Japan, and other industrializing regions of the world' (Hughes, 1983: x).

[33] 'Although the electric power systems described herein were introduced in different places and reached their plateaus of development at different times, they are related to one another by the overall model of system evolution that structures this study at the most general level' (Hughes, 1983: 14).

set academic standards, concepts and perspectives that have become ortho-
dox reference points in STS and beyond – both within and outside 'Western
society'. In fact, it is hardly possible to propose an academic study of electrification without referencing Hughes' 'landmark contribution' (Morton,
2002: 60; see also Kale, 2014; Shamir, 2013; Hausman et al., 2008; Chikowero, 2007; Coppersmith, 2004; Gugerli 1996; Nye, 1992).

This chapter presents a history of electrification in Johannesburg before
1930. Similar to Hughes' study of electrification in Chicago, London and Berlin, the historical analysis will look for emerging themes. However, rather
than applying Hughes' model of technological change to identify these
themes, the history will be developed through the analytical lenses of Trouillot's geography of management and imagination. As indicated in the introduction, technology belongs to the class of words that Trouillot calls *North
Atlantic universals*, because it creates historical standards by projecting 'the
North Atlantic experience on a universal scale'. From this perspective, rather
than operating as a descriptive category for historical analysis, Hughes'
notion of electrification advances a technological vision of history. His history of electrification comes to stand for a particular but shared perspective
on how to stage 'technology' in world history.

The geographies of management and imagination offer analytical tools
to investigate the ways in which Hughes' history of electrification in Western
society has set the terms of the debate. The electrification of Johannesburg, as
a case study located outside Hughes' bounds of 'Western society', allows us to
probe Hughes' specific calibration of the category 'electrification' and the
implications that this calibration has for his model of technological change.
For this purpose, the following chapter will map the themes that emerged in
the history of electrification in Johannesburg against Hughes' framework for
studying technological change. The purpose of this comparative discussion is
to design an alternative set of concepts for studying technological change.
Through the perspective of this alternative conceptual landscape, we can
then investigate the co-construction of STS and its object of study, and we
can consider some of the implications of STS's Collective Blind Spot on the
field of STS.

However, we must ask whether Johannesburg qualifies as a case study
for a local history of electrification alongside Chicago, Berlin and London,
which are home to 'Western society'. Johannesburg is not situated within the

bounds of this expression. Hughes does not use the term 'Western society' to insist that his model of technological change is only valid within these geographical boundaries. On the contrary, he refers to the three different styles of Berlin, Chicago and London to merge their variety into an overall model of technological change. Hughes extrapolates the conclusions drawn from the study of these three metropoles to the study of technological change more generally. The category 'Western society' is employed to designate the sites of origin of electrification, which are the sources and drivers of technological change and progress. In other words, Hughes applies 'Western society' to position the West at the beginning of electrification and to contrast it against its absence in other geographical regions. This strategy establishes the global scope of his model. The effect of plotting these trajectories of electrification is similar to that of Anderson watching the travels of STS beyond the North Atlantic. It simultaneously accords the North Atlantic experience 'analytical value and universal status' vis-à-vis its residual, which only appears on the world stage through the projected categories of this master narrative (Mamdani, 1996; Randeria, 2002: 291).

As previously mentioned, Hughes' model is regarded as a standard reference for the study of electrification. Hughes' history relies on a picture of technology transfer as the sequential distribution, dissemination, diffusion and adoption of electrification across countries after its invention and development in Thomas Alva Edison's laboratory in New Jersey. Using just the chronological measures of this picture, which in Hughes' narrative begins in 1880, South Africa would qualify as a suitable candidate. Electric lights were installed in South Africa as early as 1881. Kimberley, the city of diamonds, is often cited as having installed electric streetlights before of London. Edison electric lights were first installed in the Cape of Good Hope in 1882.

Some thirty years later, by the eve of the First World War, one of the largest electric power schemes in the world would be operating in Johannesburg. In Renfrew Christie's words, 'one of the world's most sophisticated energy systems was created [...] in a relatively undeveloped part of the British Empire' (Christie, 1984: 6). This gigantic power scheme was British and German owned and operated, but the history presented here will show that this electric power scheme was not simply the consequence of the global expansion of technology from the West to the Rest. This expansion actually played an essential role in the process of encoding the universal notion of electrification. In other words, the history of electrification in Johannesburg also shaped the history of electrification in Hughes' 'Western society'. This rather bold claim means that this book has to demonstrate that the history of electrification in Berlin, Chicago and London might have been different had it not been for the history of electrification in Johannesburg.

If we consider Johannesburg along with Berlin, Chicago and London in the early 1880s, a number of differences appear in the character of these case studies. For example, Johannesburg was only founded in 1886, whereas Berlin and London had already existed for centuries, and Chicago for decades. Johannesburg was founded as a mining camp in 1886, when gold was discovered on the Witwatersrand. The city is named 'the place of gold' in Zulu ('Egoli') and Sotho ('Gauteng'). Despite its late foundation relative to Hughes' case studies, the speed with which Johannesburg grew (driven by the gold rush) was unprecedented. Within a decade of its founding, the population of Johannesburg exceeded 100,000. From the beginning, this population was cosmopolitan in today's terms, including a wide variety of people from Southern Africa, Europe and America.

Hughes was interested in the development of regional systems. The last phase of his technological systems history ends with the maturity of the regional power systems of Chicago, Berlin and London in 1930. Therefore, strictly speaking, Hughes' unit of study is the metropolitan area in which regional power systems developed, and not their restricted city centre. Accordingly, Hughes uses the word 'Berlin' to indicate both the 'old Berlin', which encompassed 'only about twenty-nine square miles' and the 'so-called Greater Berlin, radiating ten miles from the city center', where its industries were located (Hughes, 1983: 176–7). Chicago also only acquired regional system status when it connected its power stations to adjacent municipalities (Hughes, 1983: 204).[34]

---

34 'The creation of an all-embracing system of electric light and power for Chicago was Insull's principal objective for almost two decades. In the end he reached out beyond the city and interconnected the Chicago system with suburban companies and then linked these with neighboring municipalities. The scope of the system became regional' (Hughes, 1983: 204).

For the metropolitan areas in Hughes' account, maturity was achieved by developing these regional systems in interaction with the political constituencies of municipalities and other stakeholders. Furthermore, they expanded incrementally through successive interconnection. In Johannesburg, on the contrary, a gigantic regional scheme of electrification was constructed within only a few years, just before the First World War. This scheme was developed outside the scope and powers of local municipalities or other local stakeholder groups – in fact, Johannesburg only established a town council in 1897. Electricity for municipal purposes accounted for barely 1 % of this power scheme's total electricity production in 1914.

To follow in Hughes' footsteps, the unit of analysis for this study is the wider metropolitan area of Johannesburg. Therefore, the designation '*Johannesburg*' henceforth refers to the broader area in which the regional power system was established. This area surrounding the city of Johannesburg is often referred to as the *Greater Johannesburg Metropolitan Area*, or the *Witwatersrand*. The Witwatersrand includes the areas referred to as East Rand and West Rand, which together span a low mountain range that radiates about 40 miles. For methodological reasons, therefore, Johannesburg and the Witwatersrand will be used interchangeably in this study. In contrast, the municipality of Johannesburg will be used to explicitly designate the city centre.

Within a decade, 'the gold-reef city' of Johannesburg 'had sprouted up on the veld of the southern Transvaal like an exotic mushroom' (1895) (Wheatcroft, 1985: 1). It was located in a territory referred to at the time as the *South African Republic* or the *Transvaal Republic*. This country, however, should not be confused with the contemporary *Republic of South Africa*. The South African Republic designated a geographical territory in Southern Africa that achieved independence from Great Britain in 1884. It retained this independent status until after the second Anglo-Boer War in 1901, when it again became a British colony. A decade later, in 1910, this colony was incorporated into the newly formed *Union of South Africa*, a dominion of the British Empire. The new country merged the two British Colonies (the Cape Colony and the Natal Colony), the two Boer Republics (the Orange Free State and the South African or Transvaal Republic) 'and one or two African protectorates or kingdoms' (Wheatcroft, 1985: 1). At the time of the beginnings of electrification in Johannesburg, therefore, 'South Africa was not

indeed a "country" in any ordinary sense, scarcely even a geographical entity'
(Wheatcroft, 1985: 1).

Hughes' Model of Technological Change

Thomas Hughes' *Networks of Power* tells the story of 'the growth of the electrical power system, seen as a long process of technological innovation and
development' (Barnes, 1994: 309). Hughes' study of electrification introduces
a systems approach to technological change.[35] This approach is framed by
the model of a dynamic technological network and its environment. The network is shaped by internal and external forces as well as by contingencies.
Hughes' key argument is that 'technological systems are both socially constructed and society shaping' (Hughes, 1983: 51). His model proposes the
technical system as a unit of study, the *system builder* as agent of change and
the *reverse salient* as a key mechanism of growth.[36] Technological systems
pursue a common *system goal*. System builders work at achieving this goal.

> One of the primary characteristics of a system builder is the ability to construct or to
> force unity from diversity, centralization in the face of pluralism, and coherence
> from chaos. This construction often involves the destruction of alternative systems.
> System builders in their constructive activity are like 'heterogeneous engineers'
> (Hughes, 1987: 52).

Consequently, Hughes' conception of technological change maps the evolution of a sociotechnical system from its formative period until it has reached
'maturity' and 'stability'. Hughes' study presupposes that 'the change in configuration of electric power systems during the fifty years from 1880 to 1930
constitute the formative years of the history of electric supply systems'

---

35 'The rationale for undertaking this study of electric power systems was the assumption that the history of all large-scale technology – not only power systems – can be
studied effectively as a history of systems' (Hughes, 1983: 7).
36 'System builders presiding over growth looked for, and corrected, reverse salient in
various parts of the embracing system, both technical and organizational' (Hughes, 1998:
231).

(Hughes, 1983: 1).[37] A further assumption is that the history of electric powerer systems spreads beyond national borders. For this reason, his study is comparative and focuses on three 'representative power systems from different regions for different phases of the history' (Hughes, 1983: 14). Hughes' traces the dissemination of the electric light and central power stations as well as responses to technical developments in the cities of Chicago, Berlin and London. Ending in 1930, his book covers a period of 50 years, during which the regional power systems had 'matured' (Hughes, 1982: 461). The agents engaged in, and the factors influencing, this process comprise engineers, consultants, consulting companies, legislation, local governments, etc. Although these agents and factors represent a myriad of individual, political and industrial interests in America, Germany and England, Hughes' key statement is that (despite this complexity) these histories of electrification all have something in common. To achieve this causal commonality, he coins the phrase 'sociotechnical system', which is defined as an encompassing notion for the area of complexity in question. In this way, respective local differences in relative influence on the history of electrification can be subsumed and a new order can be achieved.

Hughes' model presents an evolutionary framework that extends over time and space. On the one hand, he identifies five phases in the evolution of the sociotechnical system of electrification. The five phases are distinguished by 'themes', or 'subthemes', that emerge in the history of electrification. These serve as narrative elements for recounting the story of electrifying Western society. Hughes' five phases in the development of electrical power systems are shaped by 'essential characteristics' that introduce a number of 'key themes' that, according to Hughes, might be viewed as structuring the development of large technological systems more generally. These phases

---

37   'The half-century from 1880 to 1930 constituted the formative years of the history of electric supply systems, and from a study of these years one can perceive the ordering, integrating, coordinating, and systematizing nature of modern human societies. Electric power systems demanded of their designers, operators, and managers a feel for the purposeful manipulation of things, intellect for the rational analysis of their nature and dynamics, and an ability to deal with the messy economic, political and social vitality of the production systems that embody the complex objectives of modern men and women' (Hughes, 1983: 1).

include: 1) invention and development, 2) technology transfer, 3) system growth, 4) technological momentum and 5) maturity. Rather than paraphrasing Hughes' descriptions of these phases, they are quoted in the table below (Table 2).

On the other hand, although not theorized explicitly, Hughes distinguishes three spatial dimensions (local, universal and regional) in the evolution of light and power systems: 1) Edison's *local* (direct-current) system, characterized by homogeneity of supply and load, with similar generators and components; 2) the *universal* system, characterized by heterogeneity of load, with different generators and components; and 3) *regional* systems (or utilities) of the 1920s, characterized by increased heterogeneity, different kinds of turbines, high voltage systems and energy sources. These three spatial dimensions, in turn, come into play through the spread of material technology.

Hughes' book contains fifteen chapters. He first describes the invention and development of Edison's electric light system from 1878 to 1882. Then, Hughes dedicates four chapters to introducing the key concepts of his history of electrification: technology transfer, reverse salient and critical problems, conflict and resolution, and technological momentum. Thereafter, he devotes three chapters to recounting the history of electrification in Berlin, Chicago and London from the late 1880s to 1914. These case studies apply the previously introduced concepts and cover the three phases of technology transfer, system growth and technological momentum. Next, a chapter is devoted to the First World War. The last chapters are concerned with the maturity of regional planned systems and discuss the notions of technological culture and style.

A quote from Hughes' introduction to *Networks of Power* shall set the tone for the following history of electrification in Johannesburg.

> As a historian traditionally trained, I am reluctant to suggest a definitive model for the evolution of electric power systems. Nevertheless, I have proposed a loosely structured model because the history I explored was mostly untouched, and I want to provide some landmarks by which other historians can chart their explorations. I expect my findings to be revised, my map to be redrawn, and my themes to be redefined as the archives are explored far more thoroughly in the future (Hughes, 1983: x).

Hughes' description of the phases

I   In the *first phase*, the invention and development of a system are considered. The professionals playing a predominant role during this phase are inventor-entrepreneurs, who differ from ordinary inventors in that the former preside over a process which extends from the inventive idea through development to the time when the invented system is ready to be used. Engineers, managers, and financiers also are involved in this first stage, but they do not preside over the system's growth until later phases (Hughes, 1983: 14).

II   The *second phase* of the model directs attention to the process of technology transfer from one region and society to another. The transfer of the Edison electric system from New York City to Berlin and London is a case in point. The sites are specific, but general observations about the transfer process can be made. During this phase the agents of change are numerous; they include inventors, entrepreneurs, organizers of enterprises, and financiers (Hughes, 1983: 14).

III The essential characteristic of the *third phase* of the model is system growth. As noted earlier, the historian is responsible for analyzing growth and analyzing the growth of systems is a particularly interesting and difficult challenge. The method of growth analysis used in this study involves reverse salient and critical problems. Because the study unit is a system, the historian finds reverse salient arising in the dynamics of the system during the uneven growth of its components and hence of the overall network. In labelling such areas of imbalance 'reverse salient', the author has borrowed from military historians, who delineate those sections of an advancing line, or front, that have fallen back as 'reverse salients'. [...] In the case of a technological system, inventors, engineers, and other professionals dedicate their creative and constructive powers to correcting reverse salient so that the system can function optimally and fulfill system goals (Hughes, 1983: 14).

IV   As a system grows, it acquires momentum. The *fourth phase* of the system model is characterized by substantial momentum. A system with substantial momentum has mass, velocity, and direction. In the case of technological systems, as defined in this study, the mass consists of machines, devices, structures, and other physical artifacts in which considerable capital has been invested. The momentum also arises from the involvement of persons whose professional skills are particularly applicable to the system. Business concerns, government agencies, professional societies, educational institutions, and other organizations that shape and are shaped by the technical core of the system also add to the momentum. Taken together, the organizations involved in the system can be spoken of as the system's culture. [...] A system usually has a direction, or goals (Hughes, 1983: 15).

V   The *last phase* of system history delineated by this study is characterized by a qualitative change in the nature of the reverse salient and by the rise of financiers and consulting engineers to pre-eminence as problem solvers. Managers played the leading role during the phase characterized by an increase in momentum. In the newer phase, which involved planned and evolving regional systems, major reverse salients became essentially problems of funding extremely large regional systems and clearing political and legislative ground. Financiers and associated consulting engineers responded effectively to problems of this kind and scale. The phase was also characterized by an increased capability on the part of engineers and managers, especially consulting engineers and managers, to plan new systems and the growth of old ones (Hughes, 1983: 17).

Table 2: The five phases in the development of electrical power systems in Hughes' *Networks of Power: Electrification in Western Society, 1880–1930.*

Hughes' landmarks are gratefully adopted as a conceptual point of departure for exploring the history of electrification in Johannesburg. Nevertheless, we can read Hughes' introductory words as an invitation to continually redraw his map and redefine his themes for understanding technological change.

3.2 Electrification in Johannesburg

Hughes' history of electrification in Western society assumes that the fifty years from 1880 to 1930 constitute the formative years in the history of electrification. Histories of electrification in South Africa usually start with the first electric lights in Kimberley (1881) and Cape Town (1882) and then quickly move on to the establishment of the first regional power scheme in Johannesburg between 1905 and 1914 (e.g. Troost & Norman, 1969; Christie, 1984; Marquard, 2006; Gentle, 2008). The intermediary years have not yet received much attention in the literature.

The history of electrification in Johannesburg until 1930 can be divided into five phases. Between 1882 and 1894, the first electric lights were installed and the first private power generators were set up at the mines of Johannesburg. The first two central power stations were built during the subsequent phase (1894 to 1905) to supply power to gold mines. These power stations were registered as limited companies in London and were owned by a British company and a German bank, respectively. During the next phase (1905 to 1914), a gigantic regional power supply scheme was established within only a few years. This scheme determined the parameters for the future course of the electrification of Johannesburg and South Africa at large. It was established by a foreign-owned private company, the Victoria Falls and Transvaal Power Company (VFTPC), and its subsidiary, the Rand Mines Power Supply Company. The VFTPC was owned by the British South Africa Company (BSAC) and a consortium of German banks. The BSAC was a Royal Chartered Company with a vast set of political, administrative, mercantile and martial rights, established by the imperialist Cecil Rhodes. In the final phase (1914 to 1930), electrification was considered from a national perspective, and national legislation and regulatory institutions were created, essentially in response to the foreign-owned regional supply system in the Witwatersrand.

However, this is only four phases – indeed, the first phase has been left out in this chronology. The first phase of electrification in Johannesburg took place before the advent of the material technology in this location. Hughes locates this phase of invention and development of the electric light and central power station in Thomas Edison's laboratory in New Jersey between 1878 and 1882, far away from Johannesburg and prior to its founding. Nevertheless, this area featured in Edison's project from the very beginning. This first phase of electrification in Johannesburg remains invisible in conventional historiography, but it appears when we investigate technology as a *North Atlantic universal.*

It is important to emphasize that this study is not a critique of Hughes' particular historiography, which accentuates certain aspects and eclipses others, and applies a selective analysis of historical sources, just like any other historical account. Hughes' book was a brilliant contribution to developing the study of technology in its day. It is only because of Hughes' meticulous analysis that this study can refer back to his history of electrification as its starting point. What follows is neither a claim for historical defects in Hughes' history nor an aspiration to better historiography. The next section also provides a selective historical account that emphasizes certain aspects and eclipses others. It simply follows a different approach and set of questions, and refers to other historical sources.

1878 to 1882: Edison the Hedgehog

Hughes locates the beginnings of the history of electrification in Western society in Thomas Alva Edison's Menlo Park laboratory in New Jersey, where Edison invented and developed the incandescent light and the central power station. From this experimental geographical site, the technology was subsequently transferred to other places in 'Western society', such as New York, Chicago, Berlin and London. Accordingly, Hughes' model of technological change distinguishes between a first phase of invention and development and a second phase of technology transfer.

The first records of Edison lights in South Africa date back to May 1882, when Edison incandescent lights were fitted in the Hall of the Good Hope Lodge at Cape Town. The date of their arrival appears to fit well with the advent of Hughes' second phase of technology transfer. In Hughes' model of technology transfer, the challenge to the historian of electrification is to map the geographical transfer of electric light and power station technology from Edison's laboratory in New Jersey to Johannesburg, South Africa. This mapping accords with Trouillot's notion of the geography of management. Here we are interested in understanding the procedures and institutions that Edison implemented to develop his electric light and central power station technology. In this view, adding the case of Johannesburg might be seen as expanding Hughes' case studies to regions outside 'the West'. Hughes' chronology would remain intact, the first phase of invention and development would remain confined to the geographic radius of Edison's activities in New Jersey and New York. In this model, South Africa only enters the history of electrification with the material advent of Edison incandescent lights in 1882.

However, if we consider electrification as a *North Atlantic universal*, the historian of electrification is challenged to also map Trouillot's related geography of imagination. As the following pages will show, South Africa already appears on this map before the material arrival of Edison's electric light and central power station technology. Edison's vision of the electric light and central power station project from the very beginning included the area known today as Johannesburg. The idea of technology transfer at a global scale was not a sequel to but rather a driving force of the project's invention and development. Hughes' separation of the processes of 'invention and development' and 'technology transfer' in time and in place, dissolves in this picture. For this reason, this chapter will consider the relationship of the geographies of management and imagination of Edison's electric light and central power station technology, before its material arrival in South Africa.

Invention and Development: Hughes' Western Origins of Electrification

In chronological terms, Hughes' history of electrification begins in the fall of 1878, when funds for Edison's electric lighting project had been secured, laboratory equipment was purchased, and Edison 'employed additional men whose talents were particularly well suited for the project' (Hughes, 1983: 23). Concurrently, Edison established the *Edison Electric Light Company* as a 'patent-holding enterprise' in November 1878 (Hughes, 1983: 23). Hughes considers the need to 'acquire funds for additional laboratory equipment' as 'a major reason' for its establishment (Hughes, 1983: 23). A 'broad array of expensive machine tools, chemical apparatus, library resources, scientific instruments and electrical equipment' were acquired (Hughes, 1983: 23).

Hughes states that Edison sought assistance for dealing with the 'economic, legal and legislative factors' involved in his electric light and central power station project. In particular, Edison's attorney, Grosvenor Lowrey 'guided Edison in matters involving Wall Street, New York City politicians, and patent applications' and helped him 'to fulfill his objectives as an inventor-entrepreneur' (Hughes, 1983: 29). Indeed, Edison describes Lowrey as 'one of those who persuaded Edison to turn to electric lighting' (Hughes, 1983: 30). However, Hughes insists that despite this assistance, Edison 'played a prominent role in the financial and political scenarios concerning his inventions' (Hughes, 1983: 29). With Lowrey's assistance, Edison founded several companies between 1878 and 1882 and filed a large number of patents. Hughes also mentions the assistance of the company Drexel, Morgan & Co. of New York in the establishment of these Edison companies. In Hughes' history of electrification, Drexel, Morgan & Co. assume the role of '[disposing] of Edison's inventions in England and Europe' (Hughes, 1983: 48) and their activities are therefore seen as part of Hughes' phase of technology transfer.

> Grosvenor P. Lowrey promoted the Edison enterprises not only in the United States but in England and on the Continent as well. His activities provide an outstanding example of modes of technology transfer. Lowrey's associates, members of the great banking and investment house of Drexel, Morgan and Company, had the financial resources, the foreign contacts, and the organizational wherewithal to move technology across national boundaries. From the start of the Edison electric lighting project, Lowey anticipated the business that could be developed abroad. In October 1878 he told Edison that the way to fulfill his dream of building a working laboratory "such as the world needed and had never seen" was to sell patents, including foreign ones. It was not unusual in the 1870s for Americans to look abroad for financing and for a major market for their technology. Before the invention of his electric lighting system, for instance, Edison had promoters representing his telegraph and telephone parents in England and on the Continent. (Hughes, 1983: 47–8).

By Hughes' definition, the phase of technology transfer (and Drexel, Morgan & Co.'s involvement), set in *after* Edison's electric light and central power station technology had been developed. However, the renowned financial institution Drexel, Morgan & Co. was already involved in the project much earlier. Edison's negotiations for legal agreements with Drexel, Morgan & Co. *preceded* his electric light and central power station project – indeed, Drexel, Morgan & Co. made his project possible in the first place. Edison entered into a contractual agreement with Drexel, Morgan & Co. to exchange control over Edison's patents (that were to result from his electric light experiments) for financial support.[38] The Edison Electric Light Company was established in New York in November 1878 with their assistance.

The early involvement of Drexel, Morgan & Co. in Edison's electric light and central power station project is significant. The founding documents for the Edison Electric Light Company (November 1878), the corresponding agreements between Edison and Drexel, Morgan & Co. (31 December 1878) and between Edison and the Edison Electric Light Company (15 November 1878) record the objectives of this involvement. These involved securing patent rights, establishing companies and organizing exhibitions at a global scale. Edison's vision of an electric light and central power station project from the very beginning aimed at a global empire, including territories that would later become part of South Africa. By June 1882, a separate company had already been established to represent Edison's business interests in this territory, the Edison Indian and Colonial Electric Company, Ltd. This company predates Edison's demonstration of central power station technology at Pearl Street Power Station in New York in September 1882. In other words, Edison had secured rights and privileges in territories across the globe before the technical and economic feasibility of his technology had been ascertained.

---

38 'Financiers J.P. Morgan and Cornelius Vanderbilt arranged a three hundred thousand dollar investment in the firm, on the condition that Edison assign all patent rights on new electrical inventions to the company for a period of five years' (Sanford, 1989: 18).

The World of Edison Patents and Companies: Countries, Colonies and Dominions

The purpose of the Edison Electric Light Company was to 'fund Edison's invention, research and development projects and to bring a return on his investment through the sale or licensing of patents on the system *throughout the world*' (Hughes, 1883: 39, my emphasis). This company granted Edison financial support in exchange for patent rights. The agreement between Edison and the Edison Electric Light Company, dated 15 November 1878, clearly sets out the scope of the rights that Edison transferred to this company: '[…] the Company has been organized with the view of becoming the owner of and of making, using and vending and licensing others to make, use and vend within the *United States and other countries or colonies* hereinafter mentioned, all the inventions, discoveries, improvements and devices of said Edison, made or to be made, in or pertaining to Electric Lighting or relating in any way to the use of electricity for the purposes of power, or of illumination or heating […]'.[39] The subsequent agreement between Edison and the financial and investment company Drexel, Morgan & Co. on 31 December 1878, makes indirect reference to the area that today falls within the borders of the Republic of South Africa. This agreement proposed 'to aid Edison in obtaining electric lighting, power, and heating patents in Great Britain, Ireland, and *portions of the dominions* and to manage and exhibit the inventions described in the patents' (Hughes, 1983: 49, 50, my emphasis).

The Edison documents contrast the notion of 'countries' against the categories of 'colonies' or the 'dominions'. In 1878, the term 'dominions' was used to designate territories belonging to the British empire. The British empire in 1878 included territories that today form part of the Republic of South Africa, such as the Cape Colony, the Colony of Natal, the Orange Free State and the Transvaal. In fact, the British had only just annexed the Transvaal Republic in 1877 – the territory in which Johannesburg was established almost a decade later – and absorbed it as a British Crown Colony into the British empire.

The term 'South Africa' appears in the records of the *Edison's Indian & Colonial Electric Company, Ltd.* in June 1882, established by Edison, Lowrey,

---

39   Edison papers, [HM780053; TAEM 28:1162], my emphasis.

and Drexel, Morgan & Co. This company was 'formed for the purpose of acquiring and using in the Empire of India, Ceylon, Australasia and *South Africa* the rights and privileges of Mr. Thomas Alva Edison, relating to the application of Electricity or Magnetism as a lighting, heating, or motive agent (except the application thereof for the purpose only of locomotion on railways or tramways or common roads, and except also the right of the Cape Government to use an installation which has been already sent out to them), and for the other objects specified in the Memorandum of Association [...]'.[40] With the founding of this Company, Edison claimed to have, for a period of five years, '[...] acquired a most important advantage in the right to patent and use in Australasia, South Africa, India, and Ceylon'.[41]

However, in June 1882, a country by the name of 'South Africa' did not exist. The expression 'South Africa' referred to an indistinct territory under siege from European imperial powers. Edison's claim for rights to letters patent in 'South Africa' designated an entire geographic region.[42] The legal and

---

40   Edison Papers, 1882: (D-82-40x); TAEM 63:32, pp.1), my emphasis.

41   Edison Papers, 1882: (D-82-40x); TAEM 63:32, pp.2). Edison appointed Edward Hibberd Johnson as agent in London and attorney to manage his letters patent in the 'Colonies': 'Letters Patent specified in Schedule A hereto and have also made the applications for other letters Patent of Letters of Registration in the said Colonies for inventions of the like-character which are specified in Schedule B hereto. Now I Thomas Alva Edison have made constituted and appointed and do hereby make constitute and appoint Edward Hibberd Johnson now residing at 59 Holborn Viaduct London my true and lawful Attorney for me and my name place and stead to negotiate with any person or persons for the sale and disposition of and to sell and dispose of for any sum of money or other consideration and on any terms and subject to such stipulations and conditions as he may think fit all my right title and interest in and to all and every of the [...] April 5 1882'. Edison Papers, 1882: [D8239ZAO; TAEM 62: 827].

42   By 1883, a prospectus of some 170 pages produced by the Edison Company for Isolated Lighting in New York City, on *'The Edison System of Incandescent Electric Lighting as Applied in Mills, Stemships, Hotels, Theatres, Residences &c.'*, provided a 'List of Edison Isolated Plants in various parts of the world' which included Canada, Germany, Russia, France, Belgium, Italy, Austria, Belgium, Finland, Holland, Spain, Cuba, Chile. For the case of the incandescent lights for the House of Assembly at Parliament in Cape Town, the prospectus lists the Cape of Good Hope, a designation used for the British Colony until the Union of South Africa in 1910.

political status of this territory was complex and contested. But despite their vagueness, the terms 'South Africa', 'colonies' and 'dominions' indicate that Edison's electric light and power station technology aimed at establishing rights and privileges across the whole world as he imagined it. Importantly, Edison (together with Lowrey and Drexel, Morgan & Co.), imagined this world scale of rights and privileges before he developed the technology. Their geography of imagination drove and shaped the design of his project.

Hughes draws a clear line between the phases of invention and development. In the former, 'an imaginary device is functioning in an imaginary environment' (Hughes, 1983: 19), whereas in the latter, 'the invention is no longer an imaginary device functioning in the inventor's mind' (Hughes, 1983: 20). South Africa is a geographical territory distant from Edison's laboratory in Menlo Park, the site that Hughes identified as the geographical centrepiece for Edison and his group of inventors and technicians. However, historical documents show that South Africa figured in the geographies of management and imagination of Edison in ways that shaped the institutions and procedures that Edison initiated during Hughes' phase of invention and development. 'South Africa' formed part of the imaginary environment for Edison's technology before it had been developed. By Hughes' own measure, this means that the territories included in the term 'South Africa' influenced the phase of invention, the first phase of his technological model.

The example of South Africa confirms that Edison's plans for electric lighting and central power stations were global in scale from the very beginning. This early global scope indicates that Edison's technology was not only devised for America and Europe; it was developed for the whole world as it existed in the imperial imagination of the time. Edison pursued this global project by managing three kinds of business: patents, companies and exhibitions. As the following sections will show, these businesses, at the time, were not simply connected to the architecture of the imperial world, but contributed to its expansion.

A Global Empire of Technology

Edison had already collaborated with agents in Europe in the late 1870s to promote his telephone and phonograph patents.[43] For his electric light project, however, he sought a different league of partners. In this decision, he was advised by his 'counsel and business and financial adviser', attorney Grosvenor P. Lowrey (Hughes, 1983: 25).[44]

Edison did not select Drexel, Morgan & Co. because Lowrey's offices were in the same building,[45] or because of the company's solvency, but because this company was excellently positioned to secure his legal rights and privileges outside of America. Drexel, Morgan & Co., with offices in London and Paris, were ideally equipped to provide the kind of support that Edison needed: 'not only the financial backing of well-established bankers but also their negotiating skills and experience with regard to international agreements and the management of patents in foreign countries' (Guagnini, 2014: 157). J.P. Morgan was a crucial figure in the 'far-reaching diffusion of

---

43   'An agreement was signed in 1877 with the Hungarian-born entrepreneur Theodore Puskas, for the sale of Edison's patents in Europe (Russia, Spain, Australia, Italy, France and Belgium); one year later Edison secured the support of Joshua Franklin Bailey, who had been Elisha Gray's agent. The two established a collaboration in a joined-up effort to promote Edison's interest in Europe, signing most of their correspondence with Edison and the Edison companies as Puskas & Bailey' (Guagnini, 2014: 157).

44   According to Misa, '[...] it was Grovenor P. Lowrey that steered Edison's lighting venture through the more complicated arena of New York City finance and politics. A leading corporate lawyer, Lowrey numbered Wells Fargo, the Baltimore & Ohio Railroad, and Western Union among his clients. He first met Edison in the mid-1860s in connection with telegraph-patent litigation for Western Union and became his attorney in 1877. After experiencing the dazzling reception given to a new system of arc lighting in Paris in 1878, Lowrey pressed Edison to focus on electric lighting. While many figures were clamouring for Edison to take up electric lighting, Lowrey arranged financing for his inventive effort from the Vanderbilt family, several Western Union officers, and Drexel, Morgan and Company. The Edison Electric Light Company, initially capitalized at $ 300,000, was Lowrey's creation' (Misa, 2004: 140).

45   This banking house of Drexel, Morgan and Co. had been established in 1871 in New York by John Pierpont Morgan and Anthony Drexel. The firm was renamed J.P. Morgan & Co. in 1895, after Drexel's death.

incandescent electric lighting that occurred through a collection of companies and worldwide investments' (Hausman et al., 2008: 76). He 'had an international perspective, had overseas experience, and "vision and imagination"' (Hausman et al., 2008: 76). Drexel, Morgan & Co.'s strong ties to Great Britain[46] were especially important for establishing patent rights and companies in the imperial world: Great Britain was the unchallenged industrial and imperial power. Edison's choice of investment bankers was a strategic move to promote Edison's technology on the global stage. Edison's partnership with Drexel, Morgan & Co. was driven by his global aspirations and cleverly fitted his project into the global political economy of the time.

In Hughes' history, Edison established a large number of companies in America with a wide variety of functions and purposes between 1878 and 1883, to promote his electric light and central power station projects, and he *subsequently* established companies to represent his overseas business interests. Accordingly, his first chapter (invention and development) describes the Edison companies founded in the United States, and selected Edison companies established abroad are considered in the subsequent chapter on technology transfer. This storyline serves to substantiate Hughes' historical account, in which Edison set up these companies to develop his technological system, which was subsequently transferred to other countries. This sequential division, however, is not consistent with the historical record.[47] Hughes' history of electrification only refers to selected Edison companies and leaves

---

[46]   Morgan had close ties to London through his father, who had become a partner in George Peabody's London investment banking house, and later established J.S. Morgan & Co (Hausman et al., 2008: 76).

[47]   For example, in his first chapter, Hughes refers to the incorporation in December 1880 of the Edison Electric Illuminating Company of New York. The purpose of this 'utility, or operating, company' – a 'licensee of the parent Edison Electric Light company' – was to build the 'central generating power station on Pearly Street in New York City' (Hughes, 1983: 39). Concurrently, though not mentioned in Hughes' book, the Edison Electric Light Company of Europe was incorporated in New York in December 1880. As regards Edison's patent rights in Europe, Hughes only mentions the later establishment of the English Electric Light Company Ltd. in London in March 1882, in the chapter on Technology Transfer (Hughes, 1983: 54).

out others. Edison companies were *concurrently* set up at home and abroad to secure patent rights at a global scale.

According to Hughes, the technological system requires companies: Edison wanted to control the manufacture of the various components that were developed for the system. Furthermore, the steady progression of 'the level of experimentation from components to laboratory-scale models of the system and then to a small, pilot-scale system' required building 'a central-station system that would both function commercially and serve as a demonstration for potential franchise purchasers' (Hughes, 1983: 38). This, in turn, required the founding of an operating company.

Hughes' book offers an illustration of the companies involved in 'Edison's manufacturing system' (Hughes, 1983: 41) (see Figure 2). The figure shows two strands of companies supervised by the Edison Electric Light Company Incorporated (31 December 1878).[48] Examples of companies include The Edison Electric Illuminating Company of New York (17 December 1880), established to build the Pearl Street Central Station, and the Edison Lamp Works, established in 1880 to operate 'the world's first incandescent electric lamp factory at Menlo Park'. Hughes' text does mention in passing an agreement that was entered into with Drexel, Morgan & Co. to manage Edison's patents in Great Britain and the dominions, only a few weeks after Edison started concentrating on the electric lighting project in the fall of 1878. However, his illustration only lists Edison companies established in America, and neglects the concurrent founding of Edison companies overseas.

Edison established several overseas companies relating to electric light and power between December 1880 and June 1882.[49] Drexel, Morgan & Co.

---

48   The Figure is attributed to Jones, History of the Consolidated Edison System, p.13, Courtesy of the Consolidated Edison Co. of New York.

49   Edison Electric Light Company of Europe, Ltd. (1880, December 23); Edison Electric Light Company of Cuba and Porto Rico and Edison Electric Light Company of Havana (1881, June 10); Spanish Colonial Light Company Ltd. (successor of Edison Electric Light Company of Cuba and Porto Rico) (1882, January); Société Electrique Edison, France (1882, February 2); Compagnie Continentale Edison, France (1882, February 2); Société Industrielle et Commerciale Edison, France (1882, February 2); (English) Electric

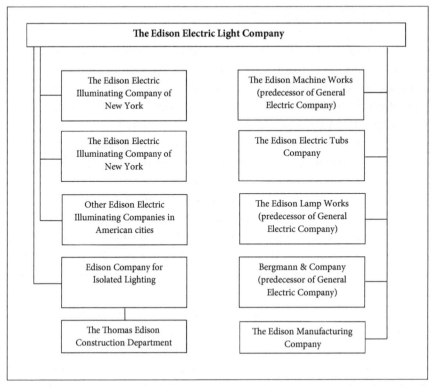

Figure 2: Abridged illustration of the companies listed by Hughes to describe Edison's manufacturing system (Hughes, 1983: 41).

managed the overseas business of Edison companies as well as the subsequent constitution in 1883 of companies in Germany, Italy, Switzerland, and Argentina. Legal arrangements to trade patent rights for equity interests also existed for Sweden, Norway, Portugal, New Zealand, New South Wales, Queensland, and Victoria in Australia (Hausman et al., 2011: 77). As a whole, the Edison companies resembled a multinational enterprise group, the legacy of which 'afforded a very strong foundation for the next phases in the international diffusion of electrical public utilities' (Hausman et al., 2011: 80).

Light Company Ltd., Great Britain (1882, March 15); Edison's Indian and Colonial Electric Company, Great Britain (1882, June 13).

The large number of Edison companies at home and abroad expresses the original vision of a global empire of rights and privileges to the electric light and central power station technology. This global strategy was determined and fixed in legal agreements *before* Edison set up his family of companies. Letters between Edison and Lowrey dated before the establishment of the Edison Electric Light Company offer glimpses of this vision. In a letter dated October 1878, Lowrey proposes to Edison the services of Drexel, Morgan & Co. This proposal is made in response to an 'enclosure from the Mexican Consul' that had been sent to him by Edison, regarding patents in Mexico. Lowrey's arguments for partnering with Drexel, Morgan & Co. in the development of the electric light are revealing of the expected benefits of this association: '[Drexel, Morgan & Co.] desire very much to control the light in all parts of Europe believing that by making one job of it, with headquarters here, the general result will be more satisfactory in every way'.[50] Edison's association with Drexel, Morgan & Co. would blend Edison's project with the power of international investment banks.

'You keep, through [Drexel, Morgan & Co.], a controlling hand upon the development of the invention on the other side so as to enforce your views and wishes, and there may arise many occasions upon which you will be very thankful that everything is managed and controlled at the corner of Wall and Broad Streets [...]'.[51]

Accordingly, a few months later, the purpose and scope of the Edison Electric Light Company of Europe was specified as follows:

The Exploitation Company has for its object the sale of the patents, the granting of licenses, in a word, the giving of value under whatever form it may be, to the said

---

50   Edison Papers, Lowrey to Edison, 10.12.1878, D7821ZBR; TAEM 18: 226, pp. 1.

51   The international dimension of patents also formed a strategic consideration for entering into business with Drexel, Morgan & Co.: 'Second. When you come to the business of disposing of the patents, it may be good financial policy not to sell outright, but to reserve interests in different places, to balance one thing against another, and to draw the largest result by allowing time. To do this rightly requires an amount of skill and power which neither you nor I possess, but which may be possessed by a great many bankers and financial people living here and in Europe'. Edison Papers, Lowrey to Edison, 10.12. 1878, D7821ZBR; TAEM 18: 226, pp. 2, 3.

patents, and this according to the conditions hereinafter stipulated. [...] The entire sale of one or several of the patents in any country of Europe, or the giving of a license for any one of these countries, cannot be made, except by express consent, given in writing, by the Light Company. [...] this contract only has reference to the following countries: 1st, France and the French Colonies (Paris with its Banlieue, Versailles included, excepted); 2nd, Belgium; 3rd Denmark; 4th, The German Empire; 5th, Austria and Hungary; 6th, Russia; 7th, Italy; 8th, Spain'.[52]

Edison patents had already been taken for the United Kingdom of Great Britain and Ireland, Portugal, Sweden and Norway.[53] Lowrey's letters reveal that the issue of the control of the rights and financial profits in overseas countries actually predated the development of any company associated with Edison's incandescent electric light and central power station. The chronological record suggests that Edison kick-started his electric light and central power station project by claiming and trading with legal rights to global entitlement and privileges. In the imperial world of the time, these claims and trades followed a geography of imagination that mapped countries, colonies and dominions. Edison's incandescent electric lamp and central power station technology and their associated institutions and procedures of control were tailored concurrently to suit this world geography.

International Exhibitions: Imperial Showcases for Technology

Drexel, Morgan & Co. acquired the rights to promote and exhibit Edison's electric light system and associated inventions in an agreement with Edison dated December 1878. As Hughes notes, this company used international exhibitions effectively (Hughes, 1983: 50).[54] He presents the International Exhibition of Electricity in Paris of 1881 as 'an informal but important exam-

---

52   Edison papers, [HM780053; TAEM 28:1162].
53   Edison papers, [HM780053; TAEM 28:1162].
54   For example, the Edison Electric Light Company (which was financed jointly by the financier and railroad tycoon Cornelius Vanderbilt and Drexel, Morgan & Co.) financed the display of the Edison system at the Paris International Electrical Exhibition from August to November 1881 (Hausman et al., 2008: 77).

ple of technology transfer' (Hughes, 1983: 51).[55] He points out the 'enthusiastic reception of Edison's incandescent lamp' at the exhibitions and its importance for the publicity and subsequent sale of Edison's technological systems and patent rights.

Hughes mentions the impression that Edison's technology made on 'scientists, engineers, inventors, financiers and entrepreneurs' (Hughes, 1983: 50) such as Emil Rathenau, Werner Siemens, Frank Sprague, Oskar von Miller and A.P. Trotter. The following year, Edison's technology was exhibited at the Crystal Palace Electrical Exhibition in London, which opened on 14 January 1882. Edison also built two prototype central power stations to exhibit his technology in New York and London in 1882. In Hughes' narrative, the Pearl Street Station in New York (September 1882) served as Edison's model central station technology, which was to be diffused 'to other American cities and to the cities of Europe' (Hughes, 1983: 47). The power station at Holborn Viaduct in London (April 1882), on the other hand, marked an important step to transfer Edison's technology to Europe.[56]

All of these technology exhibitions were held at sites of great symbolic imperial power. The Palais de l'Industrie in the Champs-Elysées, Paris, and the Crystal Palace in London were established by the two major imperial powers of the 19th century to hold international exhibitions. The Crystal Palace was built for the First World's Fairs in 1851, in Hyde Park, London, and it was later reconstructed at Sydenham Hill. The Palais de l'Industrie was established for the 1855 Exposition Universelle in Paris. The Holborn Viaduct, a richly decorated bridge built on iron girders and granite pillars, opened by Queen Victoria in 1869, was seen as one of London's Victorian civil engineering showpieces. Edison's model central power station in Pearl Street station was built in New York's lower Manhattan financial district,

---

55 'The combined effect of motivated young engineers and stimulated investors streaming away from Paris with a favorable impression of Edison and his works is an informal but important example of technology transfer' (Hughes, 1983: 51).
56 The technology of both systems was similar; 'Holborn Viaduct thus had generators of the same kind as those at New York's Pearly Street station, and its components – Edison lamps, some underground cable, and other electrical equipment – were similar to those found in the New York station' (Hughes, 1983: 55).

Wall Street, which was close to the offices of Drexel, Morgan & Co. (Sanford, 1989: 19).[57]

Hughes describes the Crystal Palace Electrical Exhibition in London and Edison's central station showpiece at Holborn Viaduct 'in the heart of London' as forming parts of a 'master scheme' of Drexel, Morgan & Co. (Hughes, 1983: 54). From the perspective of the geography of imagination, the success of this master scheme was not simply a combined result of the inherent persuasive power of Edison's technology and Drexel, Morgan & Co.'s clever overseas marketing strategy. The sites chosen to exhibit the Edison technology were significant. To exhibit Edison's technology at the Palais de l'Industrie and in the Crystal Palace was an effective step to powerfully position his electrical enterprise at the centre stage of the prevailing national and imperial power configurations of the time.

This objective is evident in the correspondence between Edison and Lowrey in October 1878. Lowrey proposed in a letter to Edison the services of Drexel, Morgan & Co.

> Before this light is introduced anywhere it must be exhibited in various places in Europe, say, for instance, London, Paris, Vienna and perhaps Berlin. These exhibitions should be under one control [...]. To sum the whole matter up, you are now in the best conceivable position which the circumstances permit. You are introduced to a new class of men who entertain the highest confidence in your ability and respect for your character. They possess all the means which may be required; (5) they live here, speak your own language, share your ideas as to what is honest and upright; are conveniently at hand to act as partners with you upon every question touching the preliminary development and subsequent management of what we all think is to be a great property. They supply precisely everything which you are

---

57   Hughes also distinguishes the phase of invention and development from the subsequent phase, technology transfer, in geographical terms. Edison invented and developed the electric light system in New York, for New York, and subsequently transferred this system to other geographical locations and socio-political contexts, where it had to be adapted to local conditions. For this reason, his chapter on invention and developments ends with the operation of the Pearl Street station in New York, and the next chapter on technology transfer recounts the circumstances for the establishment of the Holborn Viaduct electricity generating station in London, although the power station at Holborn Viaduct started operation before the power station in Pearl Street.

lacking, and *possess all the European influence or the means of getting such influence, which can ever be required.*[58]

Lowrey proposed to Edison to exhibit 'in various places in Europe', such as 'London, Paris, Vienna and perhaps Berlin', before it was introduced 'anywhere'[59] because these cities were the capitals of the great industrial nations Great Britain, France, Austria and Germany, and Europe was seen as the centre of the world. Lowrey's suggestion to Edison mirrors what Timothy Mitchell refers to as 'a particular view of geography, in which the world has a single center, Europe [...] in reference to which all other regions are to be located' (Mitchell, 2000: 7). Importantly, this staging of Europe at the centre of the world 'involves the staging of differences' in other regions (Mitchell, 2000: 26).

Maps of Technological Expansion: Europe, the Universal Unmarked

The imagined position of Europe at the centre of the world was visually presented at the International Electrical Exhibition in Paris, at the Palais de l'Industrie in 1881. The Electrical Exhibition attracted about 880,000 visitors to see the work of 1,786 exhibitors. From the eastern end of the Palais, Siemens ran an electrical tram to the symbolic Place de la Concorde (Beauachamp, 1997: 161). The venue comprised a grand nave of 182 metres, with galleries on the upper floor (Beauchamp, 1997: 161), and it was divided into two sections, one devoted to France and the other to foreign nations (Bright & Hughes, 1881).[60] Three countries occupied the major exhibition area at the

---

58   Edison Papers, Lowrey to Edison, 10.12.1878, D7821ZBR; TAEM 18: 226, my emphasis.

59   Edison Papers, Lowrey to Edison, 10.12.1878, D7821ZBR; TAEM 18: 226, pp. 2.

60   'The south side of the nave was given up almost entirely to the dynamo-electric machines, the steam and gas engines [...], the boilers, and the counter-shafting which were needed for generating the powerful currents of electricity required all over the building. The main part of this nave was divided into two equal parts, one of which was devoted to the French nation, and the other to foreign countries. The galleries and rooms on the upper story were used for miscellaneous exhibits, but were also largely utilized for illustrating the applicability of the electric light to domestic purposes' (Bright & Bright, 2012 [1898]: 591).

centre of the national section: the United States (Etats-Unis), Great Britain (Angleterre) and Germany (Allemagne).

International exhibitions in the latter part of the 19th century arranged things 'to stand for something larger' (Mitchell, 1995: 295). They attempted to reduce the world to a system of objects organized in ways that 'enabled them to evoke some larger meaning' (Mitchell, 1995: 295). These material objects also displayed industrial and imperial power. The world exhibitions of the late 19th century represented reality as an exhibit set up for 'an observing European gaze surrounded by and yet excluded from the exhibition's careful order' (Mitchell, 1995: 297). The specific arrangement of nations at the Paris Electrical Exhibition left visitors with no doubt as to the national powers at the forefront of electrical inventions. Visitors concurrently experienced displays of imperial and industrial power. This geography of countries replicated the world map of countries and colonies, as already encountered in the foundational records of the Edison companies. The exhibition imprinted in its visitors images of the Here of Europe and of its Elsewhere, the Colonies, or Dominions.

The Edison agreements and contracts between 1878 and 1882 pertaining to his electric light and central power station project replicate the geography of imagination of a world divided into 'countries' and 'colonies'.[61] Other categories employed in these documents to describe the geographical scope of the Edison patent rights include 'continental Europe'; the British 'dominions' and 'tropical countries'; 'Europe'; 'the Continent'; and the 'Colonies'. These categories offer indications about the underlying images of the world, its geopolitical boundaries and regions, and its business opportunities. They map the geographical contours of an imperial world. Edison's electric light system and its associated companies came into being in a historical era of imperial powers that were struggling for territorial control of colonies. Parallel to establishing the Edison companies in Britain and France (responsible for continental Europe), companies were formed for South American

---

61   The geographical scope of the rights that Edison transferred to the Edison Electric Light Company in November of 1878 is specified as 'the United States and other countries or colonies'. Agreement between Edison and the Edison Electric Light Company dated 15 November 1878. Edison papers, [HM780053; TAEM 28:1162].

countries and for the British, Spanish and French colonies. The resulting geopolitical map of companies reflects the imagined world geography of Edison, his lawyer and his investment banking house. This map followed in the footsteps of imperial authority and envisioned nothing less than a global empire of companies and patent rights.[62] These maps were essential for the formation of the Edison companies across the world to secure Edison's rights and privileges. As foundational background to the development of Edison's electric light and central power stations, these imaginary maps can be seen as historical drivers for Edison's electrical technology.

The territorial categories in Edison's legal documents reveal a specific feature. In addition to being divided into 'countries' and 'colonies', the imperial world had a centre: Europe. Lowrey's correspondence to Edison in December 1878 to promote an agreement between Edison and Drexel, Morgan & Co. reveals the status of Europe as 'the universal unmarked' (Trouillot, 2003). Lowrey considers this company to possess the necessary 'European influence, or the means of getting such influence' required for Edison's project.[63] This European influence allowed Edison, Lowrey, and Drexel, Morgan & Co. to seek and proclaim rights and privileges for patents and companies in far-away places such as 'South Africa'.

'International' investment law and the electrification of imagined territories

International patent law only started to take shape when Edison developed his vision of a global empire for his technology. Europe played a leading role in the newly emerging international investment law. In 1883, the Paris Convention for the Protection of Industrial Property was held to compare and

---

62   Hughes acknowledges this global scope of the Edison project in quoting a letter by Lowrey to Edison, dated 10 October 1878: 'Lowrey promised in 1878 that the income from electric-lighting patents would be enough to fulfill one of Edison's dreams: it would "Set [him] up forever—[and] enable [him] ... to build and formally endow a working laboratory such as the world needs and has never seen."' (Hughes, 1983: 30). However, he does not further pursue the significance of these global aspirations.

63   Edison Papers, Lowrey to Edison, 10.12.1878, D7821ZBR; TAEM 18: 226, my emphasis.

discuss national and international law on industrial property. This event concluded to protect the priority right across national borders. It marks the beginning of a series of conventions that established contemporary intellectual property rights.[64]

The legal institution of Edison, Lowrey, and Drexel, Morgan & Co.'s electric light and central power station projects formed part of this emerging international investment law. Edison, Lowrey, and Drexel, Morgan & Co. shared a grandiose sense of entitlement to world technological expansion of certain rights and privileges that led to the establishment of a diverse family of companies across the world. Edison's electric light and central power station project was cleverly inserted into the broad imperial power configurations of the day, but at the regional and local level of the colonies, their particular formations were in constant flux. International investment law emerged as an attempt to tame the complex set of difficulties encountered by imperial powers in their expansion of rights and privileges across the globe.

Two historical documents illustrate the kind of problems encountered by Edison's business in territories that were contested by imperial powers. The brochure *The Edison System of Incandescent Electric Lighting* (published in 1883), containing a 'List of Edison plants in use in various parts of the world'; and letters between Edison, Samuel Insull, Alfred Ord Tate (Edison's personal secretary) and the attorney's office of Edison's Electric Light Company, Eaton & Lewis, in New York in 1889.

In May 1882, sixty electric lamps were fitted in the Hall of the Good Hope Lodge at Cape Town, which was used by the House of Assembly at the time.[65] The brochure *The Edison System of Incandescent Electric Lighting* published in book format in New York in 1883 mentions the Edison lights installed at the Cape (Edison, 1883). They are itemized in a 'List of Edison

---

64   The Paris Convention, concluded in 1883, was revised at Brussels in 1900, at Washington in 1911, at The Hague in 1925, at London in 1934, at Lisbon in 1958 and at Stockholm in 1967, and was amended in 1979. http://www.wipo.int/about-wipo/en/history.html /accessed 31.5.15.

65   However, it appears that the Brush arc lamps were initially considered as more viable technological option. The South African 'Brush' Electric Light and Power Company, Ltd. was founded in London in 1882 (with a capital of 100,000 pounds) but was shut down only three years later.

plants in use in various parts of the world'. The table lists the city and country in which these Edison plants are used, among other categories. The place of installation of electric lights in the House of Assembly is recorded as 'Cape Town' (city) and the 'Cape of Good Hope' (country) (Edison, 1883). The nomenclature used in these registers is revealing: the 'Cape of Good Hope' was not a country; it was a self-governing British colony. The Edison patent rights for this colony were managed by the Edison Indian and Colonial Electric Company. South Africa during this phase designated a geographical region and not a geopolitical territory. Edison technology appeared in the region that was later to be referred to as the country of South Africa at a time when various European empires claimed, and struggled to expand, their sovereignty over land and territories.

These ambiguous and constantly changing circumstances caused by the imperial wrangle over territory in 'South Africa' are also illustrated by an order by a company in 'Johannesburg, South African Republic' for an Edison electric power station in 1889. The request required clarification of Edison's patent rights in this place. Johannesburg, however, had only been founded as mining camp three years earlier, in 1886, after the discovery of gold. The South African Republic, in turn, had only been established in 1884, as one of two independent Boer republics (the other being the Orange Free State), when the London Convention was signed after the first Boer War of 1880–81.

The request set going an exchange of letters between Edison, Samuel Insull, Alfred Ord Tate (Edison's personal secretary) and the attorney's office of Edison's Electric Light Company, Eaton & Lewis, in New York. In a letter to Tate, Insull asks:

> I have your favour of the 7th enclosing letter from L. Oscar Browning & Co., Johannesburg, South African Republic. Do you know whether this territory is controlled by the Australasian Company? Will you please look up Mr. Edison's contracts in relation to this matter and advise me further? (Samuel Insull to A.O. Tate[66]).

---

66   Edison papers, D8943AAV: TAEM 126:718.

The Edison Indian and Colonial Electric Company (established in 1882) had been absorbed in 1886 by the Australasian Electric Light Power and Storage Company, Ltd.[67] The attorneys respond:

> We cannot ascertain for a certainty whether any patents whatever are granted in the South African Republic [...]. The presumption is that there are no patents granted in that country.[68]

As to the question of whether or not the 'South African Republic' is covered by any Edison agreement, Lewis' letter to Eaton reads:

> There are two sides to this question. It may be that the said agreement was meant to cover only the English colonies, but this is not clearly expressed. The recitals in the agreement specifically mention the English colonies, but the agreement itself speaks of South Africa without any restriction. Probably all of South Africa, including what is now known as the South African Republic (no matter what it was known as in 1883) is covered by the agreement, and belongs to the Australasian Company.[69]

Why was it difficult to ascertain whether or not Edison patents had been granted for Johannesburg? The status of this territory in the early years of electrification was contested. The South African Republic, also referred to as the Transvaal Republic,[70] had been granted independence from Britain after

---

67   The Edison's Indian and Colonial Electric Company was succeeded by the Australasian Electric Light Power and Storage Company, Ltd. in 1886. The Australian Electric Light Power and Storage Co. obtained permission to promote the Edison system of electric lighting in Australasia, Ceylon, India, and South Africa. In October 1889 Edison formally assigned his patent rights to the company. In 1891 these rights were assigned to the Brush Electrical Engineering Company. Thomas Edison Papers, Rutgers. http://edison.rutgers.edu/list.htm. /accessed on 18.11.15.

68   Eaton & Lewis, New York to Edison, 26th Dec. 1889 Edison Papers, 1889: D8941ABL: TAEM 126: 666.

69   Eaton & Lewis, New York to Edison, 26th Dec. 1889 Edison Papers, 1889: D8941ABL: TAEM 126: 666.

70   The independent republics of the Transvaal and the Orange Free State were founded by Boer populations who trekked to the interior from the British Cape Colony between 1835 and 1845. These Republics were first recognized by Great Britain but later annexed in 1877.

the first Anglo-Boer War of 1880–81. This agreement was specified in the Pretoria Convention of 1881 and the London Convention of 1884. The South African Republic was granted sovereignty but with the restriction of British suzerainty. This decree demanded that its foreign agreements had to be approved by the British Government. In other words, under the London Convention, the Boer Republics had to obtain permission from the British government for any treaty entered into with another country. The category 'foreign' also included the indigenous populations and territories in Southern Africa that had not yet been taken over by European empires.

Hughes' history claims that Edison separates the question of patent ownership from 'the question of his role he played as manager and entrepreneur of invention and development' (Hughes, 1983: 28). The case of South Africa shows that these questions were connected, because the very idea of patent rights implies travel, movement and trade. Hughes' conclusion that Edison's electric light and central power station technology was designed for a particular site and later transferred to, and adapted in, other places is not supported by the historical record.[71] As we have seen, from the very beginning, Edison established an empire of patent rights that reached across the imperial world – not just 'the Western world'. In fact, Edison's technology participated in the wrangle over territorial rights and privileges in colonial settings.

The imagined territory of 'South Africa' was subsumed into foreign investment and property rules in the emerging international investment law. Such processes, according to Trouillot, 'reorganised space for explicitly political or economic purposes' at a global scale (Trouillot, 2003: 37). For this reason, they are not captured by analysing the geography of management, which focuses on place; rather, they only become visible by applying the lens of the geography of imagination, which considers both the place and space occupied by the expression 'South Africa'.

---

71 'Thus, without articulating the intent to do so, Edison and his associates designed a site-specific technology. Undoubtedly, however, they believed that a system designed for New York City would function well in the other great cities of the *Western* world' (Hughes, 1983: 47, my emphasis).

Miles identifies a direct connection between colonialism and the expansion of 'international rules on the protection of foreign-owned property [that] initially emerged from legal arrangements amongst European nations' (Miles, 2013: 2). Edison, Lowrey, and Drexel, Morgan & Co.'s global patents and companies for the electric light and central power station technology contributed to these emerging international legal rules. However, the emerging international investment law was not simply impacted by the 'colonial encounter'; it was 'shaped by it at a fundamental level' (Miles, 2013: 2). The resulting 'universal' and 'impartial' international investment mechanisms and principles 'essentially comprised protection for investors and obligations for capital-importing states to facilitate trade and investment' (Miles, 2013: 19) and 'protected only the interests of capital-exporting states, excluding the host state from the protective sphere of investment rules' (Miles, 2013: 2). By 'broadening their application to non-European nations, foreign investment and trade protection rules became part of an array of tools used to further the political and commercial aspirations of European states' (Miles, 2013: 2). This process involved 'the calculated, often brutal, use of force, and the manipulation of legal doctrines to acquire commercial benefits' (Miles, 2013: 32).

1882 to 1894: Gold and light

[…] the history of electrification in South Africa was shaped by the energy needs of the nerve centre of industry – the gold-mining industry of the Witwatersrand
Leonard Gentle, 2009.

This second phase of electrification in Johannesburg covers the years between the first installed electric lights and plans for the first central power stations. During this phase, the Johannesburg Lighting Company (later renamed Johannesburg Gas Company) was founded and built a small gas power plant for arc street lighting. Its first electric streetlights in Johannesburg were powered in June 1892. The diamond mines at Kimberly (and later the gold mines of Johannesburg) started experimenting with electric light and electric motors for various purposes on the surface and underground.[72] Electrical equipment was shipped from Britain, America and Germany.[73] Johannesburg was established as a mining settlement after the discovery of gold on the Witwatersrand in 1886. A small gas power plant was built on President Street for arc street lighting in June 1892, five years before Johannesburg was granted a town council (in 1897). The power plant was operated by the Gas Works of the Johannesburg Lighting Company. The first consulting engineer to this company was the British Engineer J. Hubert Davies.[74] Davis was also consulted to build the first electric railway in South Africa at the Crown Reef Gold Mine, which began service in 1894.

There are few references to electric installations in Johannesburg before 1894. Nevertheless, this phase is significant because it created the prime consumer for electricity in Johannesburg for the next 50 years: the gold mining industry. Within a decade of its founding in 1886, Johannesburg ranked 'as one of the commercial centres of the world' with a 'white population' of 102,000 (Oliver et al., 1985: 435). By that time, the value of gold exports from the Transvaal had surpassed the export of diamonds. By 1894, the value of gold production in the Transvaal was estimated at £7,800,000, which amounted to more than one-fifth of the world's gold production (Hatch & Chalmers, 1895 [2013]: 5). These developments gave rise to a wave of

---

72   In 1894, a book published on *The Gold Mines of the Rand* notes the following uses of electricity under the rubric 'Electric Plant': 'On most mines the shaft plant includes a dynamo for electric lighting of underground loading stations and cross-cuts, and of head-gears, sorting-floors, tramways, etc., on the surface. In some cases also there are generators for electric transmission of power to underground pumps and winding-engines. Dynamos in shaft installations are generally run by a simple high-speed engine' (Hatch & Chalmers, 1895: 147).

73   The German company Siemens and Halske Co. had already appointed agents in Southern Africa as early as 1873 to supply telegraph equipment.

74   Davies founded the mechanical and electrical engineering firm Hubert Davies & Co. in Johannesburg in 1889. Davies was involved in establishing the South African Association of Engineers and Architects and is accredited with having presented the first technical paper before the association ('Electrical transmission of power') (*Proceedings*, 1892–1894, Vol. 1, pp. 5–13). SA Biographical Database of Southern African Science. http://www.s2a3.org.za/bio/Biograph_final.php?serial=665 /accessed 12.5.2015.

immigration to the Transvaal and '[...] thousands of fortune seekers of every description made their way to the "Golden Rand"' (Oliver et al., 1985: 435). The conditions on the Witwatersrand, however, were harsh. When the gold fields were discovered, 'mining and other supplies had to be brought by ox or mule waggon from that point at very heavy cost', as the nearest railway was 300 miles away in Kimberly (Hatch & Chalmers, 1895: 244). The particular structure of the gold mining industry that developed under these conditions preconfigured the enormous demand for power in Johannesburg in the early 20th century.

However, electricity only gradually came to play an important role in the rapidly expanding gold mining industry over these years. The increasing depths of the gold-bearing ore prompted the gold mining companies to develop new organizational arrangements and extraction techniques. An important step in this process was the amalgamation of a number of individual companies into the 'system of group administration'.[75] This term referred to the unification of individual companies into a smaller number of mining houses that displayed 'a high degree of overlapping ownership' (Oliver & Sanderson, 1985: 434). This structure saved overhead expenses and afforded the mining houses easier access to capital. The need to develop new mining techniques to extract the gold-bearing ore, in turn, led the mines to consider alternative cost-effective formulas to balance the ratio of manual labour and machinery. Deeper level mining required larger capital investments. The group system also allowed the investor 'to spread his risk by investing in the

---

[75]   The group system designates the structure of mining companies where a family of subsidiary companies amalgamated under one mining house. It determined the basic corporate structure of the gold mining industry in the Transvaal: 'By 1892 there were 95 members of the Chamber of Mines representing 59 companies on the Witwatersrand and eight members representing four companies in other districts of the Transvaal. By the end of the Boer War, more appropriately, South African War (1899–1902), there were as many as nine such groups, controlling 114 gold mines between them' (Tshitereke, 2006: 32). Fraser describes this system as follows: '[...] the parent or controlling company is a mining-finance house which usually has ample financial, technical and administrative resources, as well as very considerable mining knowledge and experience. It is a shareholder in the individual mining companies and is represented on the boards of these companies by its nominees who ensure that control is maintained' (Fraser, 1975: 165).

shares of a group, which had a great portfolio of mines, rather than speculate in individual companies himself (Graham, 1996: 7). By 1894, this system allowed the gold mining companies to further develop their operations into deep-level mining.[76]

Two groups or mining houses played a pioneering role in these developments and achieved powerful positions in gold mining on the Rand: Wernher, Beit and Eckstein, and Consolidated Gold Fields. Both of these groups later played important roles in developing the gigantic power scheme that was built in Johannesburg before the First World War. The Wernher, Beit & Company was established in 1890 as a private finance and investment company, with headquarters in London. It reconstituted a company headquartered in Paris that had been founded in 1871 by the diamond merchant, Jules Porges. Jules Porges & Company sent its representatives Julius Wernher and Alfred Beit to Kimberly in 1873. Its Johannesburg business was reorganized into the firm H. Eckstein & Co. in 1889. Wernher, Beit & Company entertained close associations with financiers in Europe, such as N.M. Rothschild and Sons, and had access to capital markets in Europe.

In 1893, Alfred Beit, Hermann Eckstein and Julius Wernher pioneered the system of Group Administration by forming the Rand Mines Company Ltd., registered in the Transvaal in 1893 (Fraser, 1975: 165). This mining house was later referred to as the Corner House, and by the end of the century, it extracted about half of the gold produced in the Transvaal. Up until 1902, all of the chairmen of the Transvaal Chamber of Mines (which was formed in 1889 to represent the interests of the gold mining companies) were employees of this mining house. The Rand Mines company became a powerful force in Johannesburg and diversified its operations. For example, Eckstein was involved in the establishment of the National Bank of South Africa and the Pretoria Portland Cement Company Ltd., and in the real

---

[76] 'At the present moment the Transvaal is undoubtedly one of the most interesting countries in the world. The enormous wealth which lies buried in the Witwatersrand Gold Fields has attracted capital, enterprise, and talent, the three factors essential to the proper development of a country. With marvellous rapidity a great mining industry has sprung up, the ultimate limits of which it is difficult to assign; and the recognition of the fact that it will be possible to work the ore-deposits down to great depths, is giving rise to engineering problems of the greatest moment' (Hatch & Chalmers, 1895: v).

estate business in the Transvaal. From the beginning, the Rand Mines company recruited American engineers, and they later formalized this recruitment process through the services of the Exploration Company in London (Fraser, 1975).

The gold mining industry depended on manual labour, which was performed by 'natives', who numbered approximately 40,000 in the goldfields of the Witwatersrand in 1894 (Hatch & Chalmers, 1895: 257). The mining practices carried out by 'native labourers' on the Witwatersrand in 1895 were described as follows:

> The classes of work for which [natives] are mainly engaged are: in the mines – hand-drilling, shovelling, filling, tramming, also assisting machine drillmen, track layers, timbermen, etc.; on the surface – landing, dumping and filling trucks, tramming, ore-sorting, stoking and assisting enginemen, carrying coal, lumber, etc., pick and shovel work, assisting millmen, filling and emptying tailing vats, and generally all work carried on under strict supervision (Hatch & Chalmers, 1895: 253).

The ratio of 'native labour' to ton of ore extracted was an important indicator of the economic performance of gold mining companies: 'The native labour employed in hand drilling and in handling rock in stopes varies from $1\frac{1}{4}$ to $1\frac{1}{2}$ Kaffirs per ton of ore' (Hatch & Chalmers, 1895: 131).[77]

The category of 'the native' will be examined in more detail in the next chapter, but it is important to note the encoding of this category in Johannesburg in these early years of gold mining and electrification. The populations subsumed under this category played a central role in the early experimentation with the use of electricity in the gold mines, as illustrated by the following quote.

> Electricians object to handle over 200 volts underground, although in many instances currents of 500 and even 700 volts are now employed. Where the shafts are well timbered and dry no especial difficulty obtains in laying and protecting mains,

---

[77] 'Although mules are employed underground in some mines, for instance at the City and Suburban, the labour of tramming generally falls on Kaffirs, one or two being necessary for each truck according to its size […]. A white man or a native at each loading station is responsible for the tally of trucks according as they come from one reef or another, or contain waste rock' (Hatch & Chalmers, 1895: 132).

except where a malicious or mischievous 'boy' may attempt to cut or destroy them, in which case, if the higher voltages are used, the same boy does not usually repeat the experiment (Hatch & Chalmers, 1895: 171).

1894 to 1905: Power Stations for the Gold Mines

There are at present over 9,000 white employees at the mines, receiving wages amounting to annually over 9,000,000 (dollars), and 70,000 Kafirs, receiving in annual wages nearly 12,500,000 (dollars)
John Hays Hammond, 'South Africa and its Future', 1897.

Rough guesses place the number of natives at from two to ten millions, but, as a matter of fact, no one knows even approximately their number. This lack of information is due to the roving propensities of the natives. Here to-day, there to-morrow, it would take a mightier hunter than even the famed Selous to hunt them all down
Edgar Mels, formerly Editor of the Johannesburg Daily News on 'The South African Native', 1900.

During this phase of electrification in Johannesburg, the first three central power stations were constructed. All of these power stations were built to supply electric power to gold mines. The two first central power stations started running as privately owned companies in 1895 and 1898, financed by German and British capital respectively. Both companies, the Rand Central Electric Works Ltd. (RCEW) and the General Electric Power Company Ltd. (GEPC), were registered in London. Table 3 presents key parameters of these two companies. Both central power stations would be replaced by the larger stations of a gigantic power scheme before the beginning of the First World War. Nevertheless, they determined the sites for this expansion and also established the major customers for electricity supply in Johannesburg: the gold mines. In many ways, these foreign-owned and constructed first central power stations also set the parameters for future electrification in Johannesburg.

Hughes' conception of 'technology transfer' does not suffice to capture the defining conditions for the establishment of Johannesburg's first two power stations. The circumstances that drove the early electrification of Johannesburg are revealed when the first two power stations are considered through the lenses of the geographies of management and imagination.

The RCEW was equipped and constructed by the German company Siemens and Halske in 1895 to supply the mines of the German-owned company Goerz GmbH by way of high-voltage power lines. Goerz GmbH had been established by the Deutsche Bank in 1893. The second power station, the GEPC, was established in 1898 as subsidiary of Cecil Rhodes' Consolidated Gold Fields Group to supply electric power to six of its gold mining companies. Starting in 1902, the largest mining group of the Witwatersrand, the Eckstein/Rand group of mines, purchased power from the GEPC and acquired shares in the company (Christie, 1983). The GEPC sourced its equipment from a variety of British, American and German suppliers under the management of the British engineer Hubert Davies, who had also been involved as a consulting engineer for the RCEW. The Transvaal government granted concessions for the construction and running of these power plants. The first concession was obtained directly by Siemens and Halske in 1894, and ceded to the RCEW in 1895. The second concession was obtained by Simmer and Jack Mines in 1897 and later ceded to the GEPC, a subsidiary of the Consolidated Gold Fields.

A third power station was established in 1905 at East Rand Proprietary Mines. In addition, some mining houses set up their own electric power stations, such as Randfontein, East Rand Proprietary Mines and New Kleinfontein. Several collieries also set up their own power stations (Gentle, 2008: 54).

The first power stations in Johannesburg were not established to provide municipal power. They were built to supply power to gold mines of the Witwatersrand. Municipal electricity continued to be supplied by the Johannesburg Gas Company, but this company could not meet the increasing demand for electricity for lighting the city of Johannesburg, and an agreement was made to purchase power from the RCEW. By 1903, the city of Johannesburg was obtaining about half of its power from the RCEW. Johannesburg has remained the only South African city with piped gas infrastructure.

	Rand Central Electric Works (RCEW)	General Electric Power Company (GEPC)
Commissioned in:	1895	1898
First power generated in:	1897	1898
Owned by:	Deutsche Bank, Goerz GmbH	Consolidated Gold Fields Group (subsidiary)
Registered in:	London	London
Power supplied to:	Mines of Goerz GmbH	Simmer and Jack Proprietary Mines Ltd., Simmer and Jack East Ltd., Simmer and Jack West Ltd., Knights Deep Ltd., Jupiter Gold Mining Co. Ltd., Rand Victoria Mines Ltd. (from 1898) Eckstein/Rand Mines (from 1902)
Site:	Brakpan	Driehoek, near Germiston
Original concession:	Siemens and Halske, 1894	Simmer and Jack Mines, 1897
Equipment from:	Siemens and Halske	British, American and German suppliers
Construction:	Siemens and Halske	Hubert Davies

Table 3: Specifications on the Rand Central Electric Works Ltd. (RCEW) and the General Electric Power Company Ltd. (GEPC).

Siemens and Halske had been doing business in Southern Africa since the early 1860s.[78] According to Weinberger, both Siemens and Halske and AEG focused their African expansion efforts on South Africa. When the RCEW was established in 1895, Siemens and Halske founded the Siemens and Halske South African Agency, with its headquarters in Johannesburg, to reorganize their business activities in South Africa 'more systematically'. In the same year, the Deutsche Bank, in co-operation with Siemens and Halske, established the Technical and Commercial Corporation Ltd., a joint agency to promote their commodity exports (Weinberger, 1975: 59).[79] Three years

---

[78] Siemens and Halske also built the first central power station in South Africa. This power station was financed by the Cape Colonial Government and was built by the Table Bay Harbour Board in Cape Town in 1891. Siemens first operated in South Africa in 1860 through its British affiliate Siemens, Halske & Co. (from 1865: Siemens Brothers). It set up a telegraph line between Cape Town and Simon's Town.

[79] Concurrently, the Dresdner Bank, in co-operation with the A.E.G., established United Engineering Co. Ltd. for the same purpose (Weinberger, 1975: 59).

later, in August of 1898, the South African agency of Siemens and Halske was reorganized as a limited company, Siemens Limited Johannesburg.[80]

As mentioned in the previous chapter, the gold mining industry faced a crisis in the mid-1890s as the surface outcrops reached groundwater level (Christie, 1984). The usual gold mining methods no longer worked because the ore needed to be crushed and chemically dissolved in order to recover the gold. The gold-bearing stratum dipped south at an angle and the increasing depths of the gold ore and the different composition of the rock layer demanded mechanization of the gold mining process. This, in turn, required different mining equipment and processes, and increased power and labour, which, in turn, required greater capital investment. The mining companies responded to these changing conditions by amalgamating to form groups of companies. Although the overall demand for power increased in this period, generators driven by reciprocating piston steam engines installed at the mines continued to supply most of their power until 1905.

At the end of this phase, in 1905, Georg Klingenberg, an electrical engineer and later the director of the Allgemeine Elektrizitätsgesellschaft (A.E.G.) in Berlin, indicated that the gold mining industry in the Transvaal included 66 gold mines, handling 9,567,993 tons of ore, with a gold output value of 17,557,350 pounds. The power that these mines' machines needed was over 200,000 HP, most of which was provided by a steam generating plant, with only 25,310 HP generated by electrically-driven machines (Klingenberg, 1916: 167–8). These 66 companies, however, included a broad variety of corporate, management and ownership structures, and they were often registered under the legal authority of foreign sovereign powers. They operated as finance, exploration and trust companies under the auspices of far-away legal and political authorities.

---

[80]   'Robert Howe Gould, the son of an Englishman who had settled in Berlin to exploit his inventions, [...] was thus educated in Berlin and completed his training at the works of Siemens and Halske. Through its Johannesburg agency, this firm had received large contracts for electrical installations in the Transvaal, such as the large power-undertaking at Brakpan, referred to above, the electrical plant at the dynamite factory at Modderfontein, at Pilgrim's Rest, and so on. Young Gould was sent out to South Africa to work on these contracts' (Hahn, 1973: 2428).

During the decade from 1894 to 1905, the Transvaal Republic underwent several political changes. The discovery of gold in 1886 had spurred a tremendous influx of immigrants from Europe and America, which led European powers to become increasingly interested in gaining territorial control in Southern Africa. Germany had secured possessions in South West Africa, Great Britain annexed a large part of Zulu territory in 1887 and Cecil Rhodes acquired Bechuanaland after 1885. In the following years, Rhodes' British South Africa Company (BSAC) would also bring the kingdoms of the Matabele (1889) and the Barotse (1899) under British control.

In the years preceding the establishment of the Union of South Africa in 1910, a fierce battle for territorial sovereignty took place between European imperial powers and among local populations. In 1894, Johannesburg belonged to the Republic of South Africa (or the Transvaal), one of two independent Boer Republics that had been proclaimed in the London Convention of 1984 following the first Boer War in 1880–81. The Cape Colony, in contrast, belonged to the British Empire. The tensions between the British and Boer population were fostered by the famous Jameson Raid in 1895, which attempted to provoke an uprising against the Boer government of Paul Kruger, president of the Republic from 1883 to 1902. It was instigated by the British South Africa Company and others to overthrow the Boer Republic of the Transvaal. These tensions led to the Second Boer War (1899–1902), in which the two Boer Republics fought against the British. After the war, the two Boer republics were annexed and became British crown colonies. These colonies were placed under self-government in 1906, until the establishment of the Union of South Africa in 1910.[81] During this time, several battles were also fought against indigenous populations in Southern Africa. They were

---

81 The chief consulting engineer to Consolidated Gold Fields, the American John Hays Hammond, sketched the following picture of 'South Africa' in 1897: 'The term "South Africa" designates that part of Africa extending southward from the Zambesi River [...] to Cape L'Agulhas, the southernmost promontory of the continent [...]. It embraces German West Africa; the Portuguese territories under the administration of the Mozambique Company; Rhodesia, south of the Zambesi, under the administration of the British South Africa (Chartered) Company; the republics of the Transvaal and Orange Free State; the British Crown Colonies of the Cape, Natal, Zululand, and Basutoland, and the British Protectorates of Bechuanaland and Amatongaland' (Hammond, 1897: 234).

dispossessed of their land and laws were passed to restrict their rights. The demand for labour in the Transvaal's gold mining industry increased, and the influx of foreign settlers to the South African Republic continued.

The first two power stations in Johannesburg were built during the final years of the independent Boer Republic of the Transvaal. They symbolize two features that would continue to shape the history of electrification in Johannesburg: they were built for foreign gold mining companies (Goerz GmbH, mines of the Consolidated Gold Fields group), with capital from overseas investment enterprises that represented the interests of the imperial powers of Germany and Great Britain, respectively (Deutsche Bank, Consolidated Gold Fields group). The first central power stations in Johannesburg were built in the age of the 'scramble for Africa', before the First World War, in which rival imperial European nations claimed sovereignty over African territories. During this period, European nations conquered about 90 % of the continent's territory (see Figure 3),[82] whereas in 1870 (before this period of imperial conquest), only about 10 % of the African continent was under European control.

The capacity of these first two power stations was extended after the Boer war of 1899–1902, during which the mines had been closed (Draper, 1967: 123). According to Christie, 'the demand for energy soared in the Transvaal' during the first decade of the 20th century (Christie, 1984: 12). De Beers Consolidated Mines at Kimberley commissioned a Central Station in August 1903. The technology for the power stations and electrical equipment came from a variety of suppliers and companies in America, Germany and Great Britain, but most gold mines continued to run their own power plants with reciprocating steam engines until 1910. Electric power 'was not extensively used, but such installations as existed employed high-speed vertical steam engines as prime movers' (Draper, 1967: 123).

The electrical equipment installed in Johannesburg during these years mirrors the imperial struggle of the rival European nations, Great Britain and Germany, in Southern Africa. However, an imperial struggle for industrial monopoly was also being waged between America and Europe. Towards the end of this phase, in 1903, Emil Rathenau travelled to America on behalf of

---

82   Most of this territory was conquered by Great Britain.

A.E.G. to conduct negotiations with General Electric to divide up the world electricity market. The result of these negotiations would have a critical influence on the subsequent development of electric power in South Africa. The RCEW and the GEPC power stations were decommissioned and replaced by larger power stations after the first decade of the 20th century, but the sites were retained and formed the cornerstones of the gigantic regional power scheme that was built in Johannesburg in 1912. From the end of 1897, the Rand Central Electric Works transmitted electricity from the area surrounding the shafts of the Brakpan coal mine across forty five miles to the Main Reef Gold Mining Company on the Paardekraal farm, 10 miles west of Johannesburg. The first power stations demonstrated the benefits of electrifying mines and set the scene for the 1905 decision by the main group of gold mining companies to electrify the gold mines. They also confirmed the potential for profit to German and British capital and manufacturers. Their focus on supplying the mines bypassed the municipal needs of the Johannesburg area. From the very beginning, electrification was linked to the problem of labour in the mines. Because of their lasting influence on the history of electrification in Johannesburg, the first two power stations in Johannesburg require more detailed investigation. A number of major themes emerge in the histories of these two central power stations, and by extrapolation, we can regard these themes as formative of the history of electrification in Johannesburg.

Deutsche Bank: Rand Central Electric Works Ltd. (RCEW)

The Rand Central Electric Works Ltd. (RCEW) was registered under British company law. Perhaps for this reason, it is typically referred to in the literature as a British enterprise.[83] However, the RCEW was initiated and financed by a German bank, for a German gold mining company (owned by

---

[83] For example, Renfrew Christie's book on the history of electrification in South Africa mentions that the RCEW 'was controlled from London' (Christie, 1984: 30). Others mention that the RCEW had a local South African board but was administered from London and financed by British capital (http://www.eskom.co.za/sites/heritage/Pages/Rand-Central-Works-Limited-11.aspx). It was not unusual for German-owned companies to be registered in London.

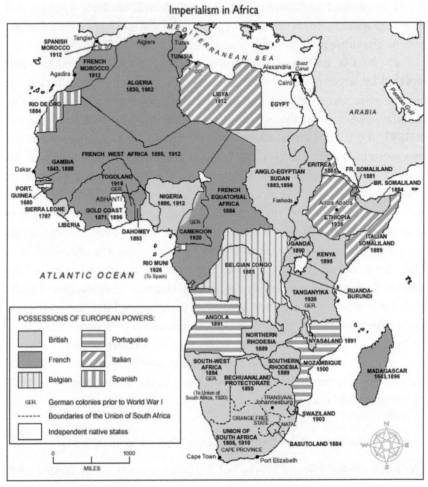

Figure 3: Map showing the 'European partition of Africa' in 1902.

the same German bank), and it was built and equipped by a German electric manufacturing company.[84] This exclusively German initiative was not a

---

[84]   Wills, for example, names The Rand Central Electric Works Ltd. as a company of the A. Goertz & Co. Ltd. (Wills, 1907 [2006]: 45). Hahn refers to the Rand Central Electric Works – 'the forerunner of the power-supply industry of the gold mines' – as one of 'many pioneer companies in Johannesburg' that were founded by Adolf Goerz (Hahn,

response to any local demand for electricity; neither did it connect in any way to local industry or municipal interests. On the contrary, the first power station in Johannesburg represented the pre-emptive strategic implementation of German industrial, financial and imperial interests overseas at the end of the 19th century.

The registration of German-owned and operated companies in London was not unusual at the time, but the silence surrounding the initiators and patrons of the RCEW has implications for the historical narrative about the development of electrification in South Africa. The RCEW is often introduced as the first electric power station to supply the municipality of Johannesburg. In fact, the RCEW only sporadically (and on request) supplied power to the municipality when demand outstripped the supply of its central gas power station. The silencing of its origins obscures the original intention for electrification in Johannesburg: to establish the German electric industry in the Transvaal (rather than for the municipality of Johannesburg) and to supply electricity to the foreign-owned, private gold mines. In 1895, two years before the RCEW started operation, the Witwatersrand had become the world's major gold producer.

The German mining engineer Adolf Goerz, who had previously worked as a consultant in Berlin, was sent to South Africa in 1888 on behalf of the Deutsche Bank. He was sent in response to a request by Eduard Lippert, a German businessman residing in the Transvaal who had obtained the dynamite concession in 1887 thanks to his 'particularly close links with President Kruger' (Jones, 1995: 7). The Deutsche Bank sent Goerz to South Africa to provide 'expert advice on the potential of the Witwatersrand goldfield' (Jones, 1995: 7). Although 'nothing came of the banking concession', Goerz's report to the Deutsche Bank resulted in the bank forming a syndicate[85] to

---

1973: 2428). Helfferich, too, considers the RCEW as resulting from the initiative of the Goerz company: 'Die im Jahre 1896 von Siemens and Halske erstellte grosse electrische Kraftanlage der Rand Central Electric Works, die gleichfalls der Initiative der Görz-Gesellschaft ihre Entstehung verdankt, ist oben (S. 132) bereits erwähnt ... (Helfferich, 1921: 132).

[85]   The syndicate also included the Berliner Handelsgesellschaft and Jacob HS Stern of Frankfurt.

'develop the Gold and diamond reserves in South Africa' by establishing A. Goerz GmbH with a capital of 100,000 pounds (Jones, 1995: 7). Following Adolf Goerz's success in acquiring directorships in various mines,[86] the syndicate was changed into the private company Ad. Goerz and Co. in 1892. This company's capital allowed Goerz to 'build up significant shareholdings and think of floating new and deep-level mines'. According to Jones, although Goerz himself had 'substantial' investments in the company, 'control remained with Deutsche Bank and its German partners' (Jones, 1995: 10). In 1897, the company registered in Pretoria, and its name was changed to A. Goerz and Co. Ltd., with 'an injection of new capital' (Jones, 1995: 11). Nonetheless, 'its discretion – and this was to be important for the future – was firmly established in London' (Jones, 1995: 11). According to Jones, four of the eight other directors were German citizens and 'the overwhelming majority of the staff came from Germany or Austria' (Jones, 1995: 11).

Goerz maintained personal ties to powerful German individuals and companies. Lord Battersea, the chairman of the new A. Goerz and Co. Ltd., was married to a member of the British Rothschild banking family (Constance Rothschild, Lady Battersea, a daughter of Anthony de Rothschild) (Jones, 1995: 11). Goerz's sister was married to Georg von Siemens, the founding director, and director of the Deutsche Bank for 30 years. Georg von Siemens was a nephew of Werner von Siemens, who had founded Siemens and Halske with Johann Georg Halske in Berlin in 1847.[87]

The Deutsche Bank's investment in the first central power station in Johannesburg to supply electricity to the gold mines was no random venture opportunity; instead, it exemplifies the bank's strategic business policy. Around 1895, 20% of the foreign capital invested in South Africa was German (Craig, 1978). German businessmen played a significant role in the economic and financial life of Johannesburg at the time. The Deutsche Bank had

---

86   These mines included May Consolidated Mines, Crown Reef, the Meyer, and Princess Estate, Metropolitan Gold Mining and New Riefontein Estate and South West Africa Company (Jones, 1995: 10).

87   Georg von Siemens had previously worked at Siemens and Halske, for which he travelled to Teheran in 1868/9 to negotiate wayleaves with the government on the construction and running of the Indo-European telegraph line from London to Calcutta.

been established as a foreign trade bank in Berlin in 1870,[88] one year before the founding of the German Empire under Otto von Bismarck. Its purpose was to 'break the English supremacy in foreign trade', which had been dominated by British banking institutions (Klawitter, 2013: 78; my translation). According to Klawitter, the dream of world trade oversees inspired the founders of the bank, and the project of global expansion also complied with Bismarck's plan of informal imperialism through economic control (Klawitter, 2013: 79; my translation). The Deutsche Bank resulted from this pioneering spirit, and 'by the turn of the century the Deutsche Bank [resembled] a spider in the web of the young German industry', pulling the strings of every future technology, including 'power stations, trams, oil, light bulbs' (Klawitter, 2013: 79; my translation).

Hughes only briefly mentions the importance of the Deutsche Bank's assistance to the two leading German manufacturers: Siemens and Halske, and the A.E.G. He describes the Deutsche Bank as 'one of the leading Kreditbanken which were, in effect, a combination of commercial and investment banks, banks that have been characterized by many historians as the centres of great historical influence – almost control – during the rapid industrialization of Germany after 1871' (Hughes, 1983: 180).

The RCEW project required the purchase of land and water rights, a supply of coal, and a concession from the Transvaal government. In 1894, Siemens and Halske sent N.J. Singels from Berlin 'to study the possibility of electrical power transmission along the Witwatersrand' (Machinery, Oct. 1897: 247). Singles purchased land at Brakpan and reached an agreement with the Transvaal Coal Trust Company[89] after having tested coal from various mines. He also secured the necessary water rights and obtained a concession from the Transvaal government to pass a transmission line along the Rand. He was assisted in these activities by the British consulting engineer Hubert Davies. After these rights had been secured, the RCEW was

---

88   The initiative to set up the bank came from the banker and businessman Adelbert Delbrück together with several other German private banks.
89   The Transvaal Coal Trust Company was established in 1889 and was taken over by Ernest Oppenheimer in 1916, renamed as Rand Selection Corporation which later became a subsidiary of the Anglo-American Corporation, established in 1917.

established in March 1895 in London. The concession was ceded, and the project was contracted to Siemens and Halske, with Mr. Singels as construction supervisor. The chief engineer of the power station from Siemens and Halske later became the managing director of the company.

The concession issued to Siemens and Halske in 1894 was organized with the assistance of the German businessman Eduard Lippert. His name already appeared in connection with the Deutsche Bank's decision to send Goerz to check out the gold mines in South Africa, and it will appear again in connection with another concession that was to influence the history of electrification in South Africa: the establishment of the Victoria Falls Power Company in 1906. Incidentally, Eduard Lippert was a cousin of Alfred Beit, whose South African mining house was financially involved in the company that established and owned the second central power station of Johannesburg. Lippert was known as a 'concession hunter' who benefitted from his personal friendship with Paul Kruger (President of the South African Republic from 1883 until 1902). Lippert obtained several concessions, such as the right to manufacture and sell dynamite in the Transvaal (1887) and the right to set up a cement factory in Pretoria (1888). The roles that Lippert and John Hubert Davies played in obtaining the concessions and agreements for the first power station in Johannesburg are unspecified, but these men were mentioned on the occasion of the transaction of the Siemens and Halske concession to the RCEW on 21 December 1895, as receivers of £5,000 and £1,000 in cash, respectively.[90] The type of concession necessary to build the first power station in Johannesburg differs from Hughes' use of the terms concessions or franchises.

---

[90]   'The subscribed capital of the company is (pound) 300,000, and at the end of 1898, (pound) 216,234 had been spent in construction expenses'. Despatches from United States Consuls in Pretoria, 1898–1906. The Rand Central Electric Works. http://www.moca vo.co.uk/Despatches-From-United-States-Consuls-in-Pretoria-1898-1906-Microform/ 110030/421, pp. 134.

Consolidated Gold Fields of South Africa: Central Electric Power Company

In 1896, while the RCEW was still under construction, the Consolidated Gold Fields of South Africa Limited decided to build its own power station rather than purchase power from the RCEW. It established the GEPC as a subsidiary company to supply electricity for six mines (the Simmer and Jack Proprietary Mines Ltd., Simmer and Jack East Ltd., Simmer and Jack West Ltd., Knights Deep Ltd., Jupiter Gold Mining Co. Ltd., and Rand Victoria Mines Ltd.). The concession to build the station at the Simmer and Jack Mine in Germiston, close to the Rand Victoria Mine, was obtained in July 1897 (Christie, 1984). This concession was ceded to the GEPC, which started transmitting power in 1898.

The GEPC was financed by Consolidated Gold Fields Ltd. After 1902, shares were bought by the Eckstein/Rand Mines group, when their mines began to purchase power from the company. By 1905, the GEPC supplied power to nine mines and the town of Germiston (Christie, 1984: 22). The power station had been designed to meet the increasing need for power in the gold mining process. Reciprocating steam engines had been used to supply power for surface gold mining, but the mines were being dug at increasingly deeper levels, which required more power for pumping water from deep-level shafts.

At first glance, the GEPC (just like the RCEW) appears to have been owned by a mining enterprise, its holding company, the Consolidated Gold Fields. However, on further inspection, the powerful architecture of associated companies reflects a dense assortment of operations and objectives in Southern Africa. Consolidated Gold Fields was founded in 1887[91] by British citizens Cecil John Rhodes and Charles Dunell Rudd, who had previously established the diamond mining monopoly in Kimberly, the De Beers Mining Company. The Consolidated Gold Fields Group branched out into an intricate dynasty of companies that Rhodes was financially involved in, such as De Beers, the British South Africa Company and the African Concessions Syndicate. According to Newbury, the 'overall management of these financial

---

91 Gold Fields Ltd. (established in 1887) was renamed as Consolidated Gold Fields of South Africa in 1892.

strands centred, however, not on Rhodes in Cape Town or at Kimberley, but on the London and Johannesburg offices of Wernher, Beit & Company' (Newbury, 2009: 96). He argues that Rhodes' association with the founders of this company, Julius Wernher and Alfred Beit, was his most important source of funds, advice and support: 'Alfred Beit was the accounting and financial genius behind Rhodes's entrepreneurial success. More than any other institution, this gold mining and industrial house influenced the management and policies of De Beers Consolidated in ways that kept Rhodes on a more stable course in business after 1895 and enabled him to restore much of his fortune for his visionary purposes' (Newbury, 2009: 97).

The engineering company of J. Hubert Davies in Johannesburg was given a contract to supply the equipment for the GEPC. Equipment and expertise were procured from various companies in America, Great Britain, Germany and Switzerland, including the General Electric Co. of America,[92] Brown Boveri of Switzerland, and Siemens and Halske.[93] The Consolidated Gold Fields recruited American electrical engineers to run and manage the GEPC.

The history of the two first central power stations in Johannesburg exposes a powerful group of German and British personalities and companies. Viewed through the geographies of management and imagination, a number of themes may be discerned as formative subjects for this phase of electrification: gold mining, empire, national competition, companies, concessions and 'the native'.

---

[92]   'General Electric had a paid company representative in the Transvaal from 1894 onward' (Hausman et al., 2008: 89).

[93]   The driving engines for the generators were made by the Allis Co. of Milwaukee, USA. The generators were made by Brown Boveri and Co. of Baden, Switzerland. The flywheels were supplied by Yates and Thom of Blackburn, England. The exciter engines were made by Bumstead and Chandler of Hednesford, England and the exciter generators by the Electrical Construction Co. of Wolverhampton, England. http://heritage.eskom.co.za/driehoek/driehoek.htm / accessed 1.7.15.

1905 to 1914: A Gigantic Power Scheme for the Gold Mines

South African electricity supply in 1905 was insignificant in world terms, but within a decade SA would have 'the largest power-works in the world', supplying 'labour-saving' machinery in the gold-mines of the Witwatersrand.
Electricity, Industry and Class in South Africa, Renfrew Christie, 1984.

A unique position is occupied by the Victoria Falls and Transvaal Power Company in South Africa; this has existed for nearly four years, and has reached an output of half a milliard (500,000,000) kw.-hours yearly [...]. It is worth while, therefore, to go into the history of this gigantic undertaking more closely, since the unusual growth of the plants can only be appreciated with a thorough knowledge of the conditions of power consumption prevailing at the time the plant was installed.
Georg Klingenberg, Large Electric Power Stations, 1916 [1913].

Between 1905 and 1914, a gigantic scheme of connected central power stations was constructed in Johannesburg – a regional system of electricity supply, in Hughes' terms. It created 'one of the world's most sophisticated energy systems [...] in a relatively undeveloped part of the British Empire' (Christie, 1984: 6) before regional power systems were instituted in Berlin or London. It was established in 1906 with capital provided by a consortium of German banks and the British South Africa Company (BSAC), and it was operated and owned by the Victoria Falls and Transvaal Power Company Ltd. (VFTPC). This regional scheme of central power stations was designed, built and equipped by the German company Allgemeine Elektrizitätsgesellschaft (A.E.G.) of Berlin. In 1907, two existing power stations, the Rand Central Electric Works (RCEW) and the General Electric Power Company (GEPC), were taken over by a company that had been established on 17 October 1906 as the Victoria Falls Power Company Ltd. (VFPC), registered in Rhodesia. In February 1908, the VFPC was renamed as Victoria Falls and Transvaal Power Company (VFTPC), and a few months later, it created a subsidiary company, registered in the Transvaal, called the Rand Mines Power Supply Company (RMPS). The two associated companies VFTPC and RMPS together were responsible for the gigantic build-up of electric power stations between

1905 and 1914. The VFTPC power scheme was designed by Georg Klingenberg, 'the engineering head of A.E.G.'s power plant design and construction
division, an engineer of international reputation and author of definite works
on power-plant design and operation' (Hughes, 1984: 197).

Within a year of its establishment, the VFTPC purchased two existing
central power stations on the Witwatersrand, the RCEW and the GEPC, and
registered in the Transvaal a subsidiary company, the Rand Mines Power
Supply Company (RMPS) in 1908, to acquire a monopoly of power generation on the Witwatersrand. In response to these developments, and at the
request of the collieries, in 1909 the Transvaal government appointed a
Commission of Inquiry into electric power supply to consider government
regulation of the industry. Based on the recommendations of the Commission, the Transvaal Power Act of 1910 was passed to provide for the establishment of the Power Undertakings Board to licence power station undertakings. The VFTPC and the Rand Mines Power Supply Company came to
operate under two separate licences granted by the Government in accordance with the Transvaal Power Act of 1910 (Hadley, 1913: 3). In effect, this
legislation served to sanction the foreign German and British monopoly in
power supply on the Witwatersrand.

The VFTPC power scheme

By 1914, the VFTPC and its subsidiary the Rand Mines Power Supply Company provided electricity to the gold mines from four central power stations
that were connected through transmission lines and overseen by a central control station at Simmerpan. This early regional scheme fixed the terms and conditions, as well as the sites and main customers, for the future development of
electrification in Johannesburg. Apart from a few local adaptations, the layout
of these power stations was similar and amounted in 1913 to a total capacity of
power installed and in progress of 176,000 kW (see Table 4).

In addition, power was distributed by another method that 'had not previously been considered on so large a scale' (Hadley, 1913: 22): six electrically-driven air compressors operated at Robinson Central air station at a
capacity of 3,500 kW each (Hadley, 1913: 5).

Name of Station	Total Capacity of Electric Generating Plant installed	Steam-driven Air Compressors installed	Extensions in progress
Brakpan	Two 3,000 kW sets	-	-
Simmer Pan	Six 3,000 kW sets	-	-
Rosherville	Five 10,000 kW sets	Six 3,500 kW machines	Three 7,000 kW steam-driven air compressors
Vereeniging	Four 10,000 kW sets	-	-
Extensions in 1913	-	-	Two 10,000 kW sets
	113,000 kW	21,000 kW	41,000 kW
Total capacity of plant installed and in progress: 176, 000 kW			

Table 4: Total capacity of power installed and in progress in the four power stations of Johannesburg in 1913 (Hadley, 1913).

In 1913, Hadley[94] characterized the VFTPC power scheme as 'a single system'[95] connected by 40,000 volt overhead transmission lines 'stretching practically the whole length of the reef' (Hadley, 1913: 5). He described the area 'over which a power supply had to be given' to lie 'within a strip about two miles broad and stretching 50 miles from east to west' (Hadley, 1913: 4). Johannesburg was situated 'about the middle of this strip', but it had its own electric plant (Hadley, 1913: 4), a gas power station.

Electricity was fed into this power scheme at four central power stations: Brakpan, Simmerpan, Rosherville and Robinson Central. At Robinson Central power station, the electricity generated at the Vereeniging power station, some 36 miles to the South of Johannesburg, connected with the system. The transmission lines ran through 'two distributing centres at Hercules to the east and Bantjes to the west' (Hadley, 1913: 6). Together, these six points composed a network to supply the 'step-down transformer stations' of

---

94   The paper *Power Supply on the Rand* was presented to the Institution of Electrical Engineers in 1913 by A.E. Hadley, who succeeded Price as general manager of the VFTPC in 1926, and later served on the Board of Directors of the BSAC.

95   'The system has been laid out so as to be operated during the development of the undertaking as a single system, but arranged that when the growth of the load made it desirable (both from the point of view of economy and also of safety of supply) it could be sectionalized without the necessity of running additional machinery' (Hadley, 1913: 9).

the mines (Hadley, 1913: 6, 29).[96] Simmerpan also operated a central control system to control and supervise the technical and safety regulations of the transmissions network, and later it continued to assume this function as national control centre. The repair, refurbishment and testing of equipment and instruments was also centralized, at a workshop and testing department at Rosherville.

In 1914, a few other power stations were providing electricity in Johannesburg, although they were hardly comparable in capacity to the supply of the VFTPC: the gas power station of Johannesburg Municipality (13.25 MW) and the East Rand Proprietary Mines (19.95 MW), Randfontein, and New Kleinfontein, but all of these stations 'eventually took additional supplies from the VFTPC' (Christie, 1984: 17).

The VFTPC built its monopoly in successive steps but at a rapid pace. In June 1905, the Rand Central Electric Works extended its existing triple-expansion engines with a 400 kW steam turbine. Plans for additional extension a year later (in December 1906) with a 2500 kW turbine were made but not executed because of VFTPC's purchase of the RCEW in February 1907 (Power Company Commission, 1910: 9–10). In 1908, the VFTPC built a new power station next to the generating station of the RCEW at Brakpan that consisted of two 3000 kW turbo generators (Power Companies Commission, 1910: 10).[97] In 1909, the power station at Simmerpan, built by the VFTPC close to the site of GEPC, was equipped with six 3 MW sets and two 11 MW sets (Troost & Norman, 1969: 178). These extensions were made to accommodate 'a complete re-design of the distribution system' (Power Companies Commission, 1910: 10). The VFTPC built two further stations, one in Rosherville in 1911 (five 9.6 MW sets) and one in Vereeniging (two 9.6 MW

96   'The electrical supply at 2,000 volts and 550 volts to the consumers' premises is effected from step-down transformer stations, which are built by the consumers, but are equipped with switchgear and transformers by the power company. There are 60 of these consumers' sub-stations connected to one system, and their individual capacity varies from 10,000 k.v.a., the normal size being 5,000 k.v.a. [...] The transformers have been supplied by Messrs. Siemens, the Allgemeine Electricitäts Gesellschaft, and the Westinghouse Company' (Hadley, 1913: 29).

97   These turbo generators were later extended by 12.5 MW sets in 1915 (Troost & Norman, 1969: 178).

sets in 1912 and two 12 MW sets in 1913), located 36 miles to the south of Johannesburg on the banks of the Vaal River, close to a colliery.[98] The total installed generating capacity of the VFTPC in 1915 (162 MW) did not increase during the First World War (Troost & Norman, 1969: 178).

The power requirements of the mines were supplied by means of electric motors that were used for winders, pumps, stamp mills, compressors, tube mills, hauling and conveying, crushers, workshops and ventilation (Rider, 1915: 613). Electrical mechanization of the labour in the gold mines directly eliminated jobs (Christie, 1984: 17). Hand tramming, the 'horizontal movement of ore underground' was replaced by electrical systems and underground haulage roads: 'The motive behind this electrical mechanisation was the displacement of black workers because of their relative scarcity "with electric locomotive traction underground tramming costs could be reduced to less than half of what they are with native labour, and a great many natives would be liberated for other work in the mine"'[99] (Christie, 1984: 17). Electricity was also used for smaller applications at the mines:

> In 1905 the Eckstein group measured twelve processes for electrification which were "mine-pumping; winding; air-compressing for drilling; crushing and sorting; stamp milling; raising of tailings; grinding in tube mills; cyanide treatment; tailings and other surface haulages; lighting; motor repair shops and grinders for the assay offices"[100] (Christie, 1984: 18).

Over the years, various engineers have commented on the ideal conditions of the Witwatersrand for a large power supply scheme. In 1941, Jacobs mentions the 'relatively compact' and 'considerable number of large consumers in a terrain devoid of natural geographical obstacles', as well as the availability

---

[98] According to Troost and Norman, 'Vereeniging was the first power station in South Africa (and possibly in the world) to be located on a coal mine and it is thus the forerunner of the giant pithead stations subsequently constructed in South Africa' (Troost & Norman, 1969: 178).

[99] Quoting Mackie, R.G. 1910. General Application of Electricity to Rand Mines Underground. *SAE* /Nov. 1910, pp. 81.

[100] Quoting Heather, H. and Robeson, A.M. 1908. On the Cost of Power at Mines of the Witwatersrand with Reference to a Proposed Supply from a Central Source. *Min. Proc. Inst. Civ. Eng.* (1906–7), Vol. IV, pp. 345–6.

and reasonable pit-head prices of coal from the nearby collieries (Jacobs, 1941: 259). Klingenberg adds that there were 'low building costs in the vicinity of a large town (JHB)' (Klingenberg, 1916 [1913]: 201) and 'exceedingly simple' 'legal conditions for the transmission of energy on the Rand by means of overhead lines and underground cables' (Klingenberg, 1916 [1913]: 240).

> The legal conditions for the transmission of energy on the Rand by means of overhead lines and underground cables were exceedingly simple. As the land in this district consists almost exclusively of mining claims, the rights of property of mine owners are restricted by legislation to the minerals underground, while the land itself is at the mine's disposal only as far as it is required for working the ore. The authorities are therefore free to grant permission to a third party to use the land for transmission lines. When the company took over the Rand Central Electric Works, it also acquired the concessions granted to that company. It was therefore protected against opposition on the part of the mines (Klingenberg, 1916 [1913]: 230), provided that their buildings were not interfered with, a condition which could easily be fulfilled (Klingenberg, 1916 [1913]: 240).

The transmission lines from the power station at Vereeniging, however, passed over private property, and therefore they were subject to other conditions, and 'difficulties arose similar to those we are accustomed to encounter in Europe' (Klingenberg, 1916 [1913]: 240). Although a concession had been purchased from the Lewis & Marks collieries (who had received wayleaves for a transmission line in 1905), construction had to wait for the investigations by the Power Companies Commission (Hadley, 1913). Accordingly, Klingenberg reports that preliminary work, such as 'the purchase of land, borings, locality of site, and questions of water supply', were settled in 1909 while 'the commencement of building operations was, however, delayed by negotiations with the authorities in connection with the necessary wayleaves' (Klingenberg, 1916 [1913]: 239).

In 1915, in a paper entitled 'Power Supply to the Mines of the Rand', Bernard Price, general manager of the VFTPC, noted 'an outstanding feature' of the VFTPC: 'that no less than 99 per cent of the business is represented by supply to the Mining Industry' (Price, 1915: vi), with only 1% supplied to other industries and municipalities (see Table 5).

Class of Consumer	No. of Consumers	Aggregate of Non-simultaneous Maximum Demands at Points of Delivery, January, 1915		Units Consumed per Annum	
		kW	Percentage of Total	Millions of Units	Percentage of Total
Mining	65	95,000	98.0	514.5	99.0
Industrial	8	1,100	1.1	2.8	0.55
Bulk supplies to Municipalities	5	900	0.9	2.3	0.45
Totals	78	97,000	100	519.6	100

Table 5: Particulars of electricity consumers (mines, industries, municipalities) supplied by the VFTPC in Johannesburg in 1915 (Price 1915: vi).

Origins and purpose of the VFTPC

The establishment of the VFTPC is typically regarded as the starting point for the history of electrification in South Africa. These beginnings are usually seen as a response to the rising need for power in the gold mines in the early 20th century. However, when the establishment of the VFTPC is considered in light of the prior history of electrification in Johannesburg, a different picture emerges. From this perspective, the development of the VFTPC is a sequel to the battle between the powers involved in the development of the first two central power stations in Johannesburg, the RCEW and the GEPC. The forces that shaped these two power stations continued to compete in the run-up to the VFTPC: (foreign-owned) gold mines; British investment companies and financial enterprises; German investment banks and their associated electrical industries and technology; British imperial mining and concession companies, especially those connected to Cecil Rhodes and Alfred Beit; and American, British and German engineers.

The establishment of the VFTPC, as indicated by its name and country of registration, was initially associated with the idea of generating power from the Victoria Falls on the Zambezi River in Rhodesia, but such a project was not realized. Therefore, the name of this company (which continued to exist until 1948) is misleading. Prior to the establishment of the VFTCP, there were numerous initiatives to establish power stations in the wider

Johannesburg area, but the VFTPC put an end to those initiatives. Important considerations for this phase of electrification are the origins of the idea of harnessing the power of the Victoria Falls and transmitting it to the gold fields of the Witwatersrand; the pioneers, negotiations and initiatives leading to its foundation; and the legislation that was subsequently created to endorse the VFTPC/A.E.G. monopoly.

Most accounts of the establishment of the VFTPC attribute the idea of generating electric power at the Victoria Falls and transmitting it to the gold mines of the Witwatersrand to the rising need for electrical power of the machines needed to work the deepening levels of gold mining on the Witwatersrand in the early 20th century. Similarly, the establishment of the African Concessions Syndicate, the subsidiary of the BSAC, is erroneously dated to the early 1900s. The African Concessions Syndicate had already been founded in 1895 to acquire the concession (from the BSAC) to generate power at the Victoria Falls. There is evidence that the idea of power generation at the Victoria Falls dates back much earlier, even before the power works at Niagara Falls, the model that inspired the idea, went into service in 1895.

George Forbes, the 'chief electrical consultant' (Hughes, 1983: 139) to the Niagara Falls power works had already been consulted on the matter of power generation and transmission at the Victoria Falls in 1894. In an article published in November 1898 in the *Journal of the Society of Arts*, Forbes reported that he had been asked 'by letter from Johannesburg whether it would be possible to transmit power from the Victoria Falls, on the Zambesi, to all the gold mines in Rhodesia, varying from 350 to 500 miles distance' (Forbes, 1898: 25). He subsequently travelled to South Africa 'to negotiate with Mr. Cecil Rhodes and Dr. Jameson for a concession'. Evidently, he considered the idea realistic and profitable. However, according to Forbes, the draft concession to the Chartered Company was interrupted by 'the Jameson raid and the Matabele rising' (Forbes, 1898: 25).

The African Concessions Syndicate Ltd. was established and registered on 4 October 1895 (Skinner, 1911: 371). According to Wills, the African Concessions Syndicate Ltd. 'was formed [...] to acquire a concession of the Victoria Falls on the Zambesi River, South Africa, for the purpose of utilizing the water power for the generation of electricity, and has obtained from the British South Africa Company a lease of the Falls for 75 years under certain conditions' (Wills, 1905: 282).[101] Its balance sheet for 1903 discloses an amount of 11,104 pounds for 'obtaining and maintaining Concession as of June 30 1902' (Wills, 1905: 283). Its consulting engineers were listed as Sir Douglas Fox and Sir Charles Metcalfe, the chairman was W.A. Wills (not W. H. Wills), and the director was Henry Wilson Fox (involved in the Matabeleland and Mashonaland uprising in 1896, and manager of the Chartered Company from 1896).

Authors differ on the origin of the idea for generating and transmitting electric power from the Victoria Falls, but they all regard its origins as being in the early 1900s and tend to mention parallel initiatives involving the A.E. G., the Dresdner Bank, the BSAC and Harper (among others) that were concerned with assessing the viability of the project. However, there is evidence that the early roots of the idea formed part of Cecil Rhodes' imperial British expansion and appropriation of territories in Mashonaland and Matabeleland. The BSAC referred to these territories as Rhodesia in 1895. The timing is no coincidence, as the idea formed part of Rhodes' various strategies to establish white settlement in Rhodesia, which, in turn, was part of his vision to create a British Empire from the Cape to Cairo (Figure 4). One key to achieving this empire was a railroad between the Cape and Cairo (Figure 5). The objective of this idea was to expand imperial British territories, and providing electricity to the gold fields was a means to this end.

> In the current year, as has been stated, the Victoria Falls have been reached by the Cape-to-Cairo Railway. That fact is fraught with vast possibilities for Rhodesian enterprise. That railway is already stimulating the development of Rhodesia, and the transformation now quickly coming to pass is one of the most impressive in the history of colonization. In the near future, however, there looms the most momentous achievement of all – namely, the harnessing of the Victoria Falls [...]. The Victoria Falls are about two and a half times as high as those of Niagara, and they are approximately twice as wide (Wills, 2006 [1907]: 264).

---

101   Both Weinberger and Klingenberg incorrectly date the founding of the African Concessions Company Ltd. (which acquired the rights to build a power station up to 250,000 PS at the Victoria Falls), subsequent to the second Anglo-Boer War although it was already established in 1895.

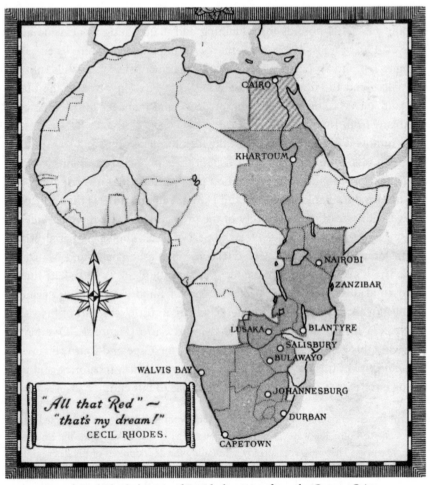

Figure 4: Map of Cecil Rhodes' vision of British dominion from the Cape to Cairo.

Various historical accounts have been put forward of the conditions surrounding the establishment of the VFTPC. Because of its importance in the
history of electrification in Johannesburg, and in South Africa, various interpretations of this historical moment are outlined below.

The first power stations of the VFTPC, and the related first regional
scheme for power generation in South Africa, were designed by the A.E.G.
engineer Georg Klingenberg, who succeeded Rathenau as head of the division for the construction of central power stations. In 1913, Klingenberg

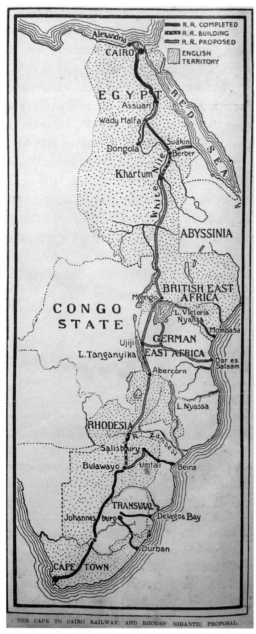

Figure 5: Map of Cecil Rhodes' plans for a railroad line
from the Cape to Cairo.

wrote a textbook that was considered to be standard literature in engineering courses for many years. It was translated into English in 1916. The two-volume book is entitled *Large Electric Power Stations: Their Design and Construction. With Examples of Existing Stations*, and it dedicates almost 100 pages to the instructive example of the VFTPC. Klingenberg provides engineering, financial and historical information on the power stations that were built and planned by the A.E.G. for the VFTPC up until 1913. His account presents the establishment of the grand centralized regional power system as a response to the rising demand for electric power at the gold mines of the Witwatersrand.[102] This view is shared by Christie, who describes the soaring demand for energy by the increasingly deepening and mechanizing of the Transvaal's gold mines in the first decade of the 20th century (Christie, 1984: 12).[103] Klingenberg claims that 'the proposal for a general power supply on the Rand was taken up almost simultaneously by the different parties involved' (Klingenberg, 1916 [1913]: 169), including the A.E.G.,[104] the

---

102   Klingenberg's description of the VFTPC begins with a review of the gold mining situation on the Rand and its power needs which at the time were supplied by means of steam plants. He maintains that these 'working conditions prevailing on the Witwatersrand made it imperative, in this instance, more than in any locality, to centralise the generation of power' (Klingenberg, 1916 [1913]: 169).

103   Christie describes the situation as follows: 'The supply of black labour was declining relative to the increased needs, at least in part because of wage reductions made during and after the Anglo-Boer war. The mine-owners' major response to this relative labour shortage was the use of more machinery. Unable to increase the gold-price, and faced by inflation, the mine-owners were forced to cut costs by mechanisation. This led to the creation and growth of the Victoria Falls and Transvaal Power Company Limited' (Christie, 1984: 12).

104   Klingenberg reports that the A.E.G., 'in connection with the German Dresdner Bank, commissioned two engineers, Loebinger and Dr. Apt, to carry out exhaustive investigations of the power demand and other conditions upon which the centralisation of a power supply depended' (Klingenberg, 1916 [1913]: 166). The date of their trip, however, is not mentioned. Based on these reports the A.E.G. developed a scheme with an output of 30,000 KW of which half was to be provided by power stations at one of the large dams of the Rand (Klingenberg, 1916 [1913]: 173).

Eckstein group[105] and the British South Africa Company.[106] However, Klingenberg's account of the negotiations leading to the grand power scheme does not provide dates to support his chronology of developments.

One set of negotiations was with the BSAC (and its subsidiary, the African Concessions Syndicate), which resulted in an agreement for a general scheme and the establishment of the Victoria Falls Power Company. The second set of negotiations took place with the Eckstein group of mines, 'the largest of the mining concerns' (Klingenberg, 1916 [1913]: 192).

The Eckstein group had issued a 'provisional contract for the supply of energy to the Eckstein group' (Klingenberg, 1916 [1913]: 192) to the British engineer W.A. Harper. At the time, the municipality of Johannesburg had commissioned a new gas engine generated power station, but because it ran into difficulties, the city continued to obtain electricity from the Rand Central Electric Works, and thereafter from the VFPC. According to Klingenberg, the contractors of the plant 'commissioned W.A. Harper, in the spring of 1908, to negotiate on their behalf' (Klingenberg, 1916 [1913]: 192). Harper developed an 'entirely new scheme' of supply to a wider area, and he negotiated with the municipality and various mining companies. He obtained a 'provisional contract for the supply of energy to the Eckstein group' (Klingenberg, 1916 [1913]: 192). In 1916, Klingenberg judged this contract to be 'the most important contract of this kind that has ever been made with a single consumer' (Klingenberg, 1916 [1913]: 193). The Eckstein Group

---

105 Klingenberg names Robeson, 'at the time chief engineer of the Eckstein group as the first to have worked out a scheme of about 70,000 HP to provide electricity and compressed air to the mines of the Eckstein group and the Consolidated Gold field' (Klingenberg, 1916 [1913]: 169). Again, the date of this scheme is not given.
106 Klingenberg declares that the British South Africa Company simultaneously, but independently 'also studied the question of power supply on the Rand, with the object of utilising the enormous water power of the Victoria Falls on the Zambesi river, which belong to them' (Klingenberg, 1916 [1913]: 173). According to him, H.W. Fox, manager of the BSAC, founded the African Concessions Syndicate in order to acquire 'the right from the British South Africa Company to erect a power station with an output of 250,000 H.P. at Victoria Falls' (Klingenberg, 1916 [1913]: 173). Since this required technical and economic investigations, 'a committee of experts' was formed (Klingenberg, 1916 [1913]: 173).

guaranteed a maximum consumption of 80,000,000 kWh per annum and agreed to replace their existing steam plant with electricity purchased from power works that were to be established under the contract. Thus, Harper needed to establish a company to finance the new power stations, and according to Klingenberg, he approached the A.E.G. 'Successful negotiations' followed between Klingenberg and the 'principals of the Eckstein group', and 'a modified contract' was signed with the VFPC, after Lord Winchester, chairman of the VFPC, 'had been successful in securing the large extra capital the company required for the new undertaking' (Klingenberg, 1916 [1913]: 194).

However, financial power or technical excellence alone still appear insufficient to explain why the Chartered Company would be prepared to concede rights to the German A.E.G. – given the rivalry between the imperial and industrial powers, Great Britain and Germany. Wernher, Beit and Company, as a leading shareholder of the Chartered Company,[107] was also represented on the board of directors of the A.E.G. after its merger with the Union Elektrizitätsgesellschaft in 1905. The London financing house Wernher, Beit and Company was one of the creators of the VFTPC, and it was given a preferential position with the establishment of the Rand Mines Power Supply Co. as a subsidiary of the VFTPC. The mines of the Eckstein group would receive electricity at a lower rate than other consumers (in 1910, 3.723 Pfennig per kWh, other customers 6.33 Pfennig) (Weinberger, 1971: 72). The magnitude of this deal cannot be overemphasized: in 1911, the Eckstein group of mines produced over 9,000,000 of the 24,000,000 tons of ore crushed on the Rand (Rickard, 1922: 235). Its employees comprised 'about 8000 whites and 56,000 natives' (Rickard, 1922: 235). The Rand Mines and the Eckstein group played an important role in the move to deep-level mining after the Anglo-Boer War and the amalgamations of the Central Rand between 1906 and 1911, which transformed the productive base of the

---

107   According to Albrecht, Alfred Beit helped Rhodes establish and finance the British South Africa Company. He became a member of its board of directors in 1889 and took over shares to the value of 34,000 pounds. Rhodes owned 75,000 pounds, De Beers 200,000 pounds and Gold Fields almost 100,000 pounds and the Exploration Company 75,000 pounds in shares (Albrecht, 2011: 60).

mining industry.[108] In 1908, during this period of transformation, Rand Mines Ltd. and the Eckstein group decided to 'change over their mines to electric driving' (Hadley, 1913: 3).

According to Bernard Price, the founding Chief Engineer (and, from 1926, General Manager) of the VFTPC,[109] the Eckstein Group stipulated (in their contract with A.P. Harper) the terms for the VFTPC to supply the Eckstein mines. First, 'a separate company should be established for supplying

---

108 The amalgamation of the gold mines of Johannesburg in the early twentieth century is described by Richartdon and van Helten: 'Both companies [Rand Mines/Eckstein] took a particularly prominent part in the merger movement which swept across the Central Rand between 1906 and 1911. The key year for the Wernher Beit companies was 1908/9. [...] no less than six major mergers in this year involved this group, out of which grew extensions [...] and a wholly new mine, the Crown Mines. These changes involved the disappearance of twenty-two separate mining concerns. In addition, in November 1909 Rand Mines Ltd. acquired the whole of the claim property of Rand Mines Deep Ltd.' (Richardson & van Helten, 1982: 330). They consider these amalgamations to have impacted on the entire region: 'This shift to predominantly deep-level mining after 1902, as exemplified in both the amalgamations of the Central Rand and the opening of the Far East Rand, had significant consequences. The productive base of the industry was transformed. Large-scale industrial organization and financial criteria became the predominant forms of operation' (Richardson & van Helten, 1982: 334).

109 Bernard Price came to South Africa in 1909 as founding Chief Engineer (and from 1926 General Manager) of the Victoria Falls and Transvaal Power Company. He had previously been employed by the consulting engineering company Merz and McLellan in London. Price maintains that he was a member of staff of Merz and McLellan in Newcatle-on-Tyne when A.W. Harper approached financial houses in London with his contract with the Eckstein group, and these 'at once called in for advice the leading firm of Consulting Engineers in the field of electric power supply, Messrs. Merz and McLellan' (Price, 1941: 270). Other authors have claimed that it was Klingenberg who persuaded Price to assume this post in South Africa (Christie, 1984: 38): 'The Germans [...] had enough sense to see that they must manage the thing through Englishmen, and they got our friend, Arthur Wright, to act as a general adviser, and, later on, Klingenberg persuaded Bernard Price (who was head of our Electrical Department), whom he had met in connection with Durham work, to go out to Africa as their chief engineer. This was the beginning of Price's long and successful career there, and of the huge undertaking which he has built up, the Victoria Falls and Transvaal Power Company' (Christie, 1984: 38, quoting from C. Merz: Charles Hesterman Merz, (TS, 1936), pp. 2.

the Group's requirements' (Price, 1941: 271). Second, the power company had to supply compressed air 'at a price which was roughly equivalent to that which would have been paid had the Mines of the Group electrified the compressors then existing on their properties and purchased electricity under the Contract for driving them' (Price, 1941: 271).[110]

This capital of £1,800,000, according to Klingenberg's declarations, included equal portions of debentures and shares. The debentures 'were again subscribed in Germany, the Deutsche Bank participating on this occasion, with the result that a share of the plant was manufactured by the Siemens Schuckertwerke' (Klingenberg, 1916 [1913]: 194). The 'idea of developing the water power of the Victoria Falls had in the meantime been abandoned' because of pressure from the coal owners, who 'had prohibited the importation of electrical energy into the Transvaal (Klingenberg, 1916 [1913]: 194). This position led the Victoria Falls Power Company Ltd. (VFPC) to change its name to the Victoria Falls and Transvaal Power Company (VFTPC) in 1908 and to register a subsidiary company in the Transvaal named Rand Mines Power Supply Company (Klingenberg, 1916 [1913]: 194).[111]

The A.E.G. was contacted to construct new power stations and to supply seven steam turbines and equipment for transmission lines, a cable network and a network of air pipes (Klingenberg, 1916 [1913]: 195). According to Klingenberg, the RMPS appointed him (Klingenberg), Arthur Wright and W.A. Harper as consulting engineers.[112] Finally, Klingenberg points out that the impending increase in the demand for power by the Eckstein group was

---

110   'This presented the Power Company with the somewhat novel and by no means easy problem of establishing a centralised scheme for the production and distribution of compressed air on a much larger scale than had previously been attempted anywhere in the world' (Price, 1941: 271).

111   Klingenberg mentions a trip to South Africa with Lord Winchester in spring 1909 to discuss the design and sites for the power stations, and claims that Lord Winchester proposed to form the Rand Mines Power Supply Company (Klingenberg, 1916 [1913]: 193).

112   Klingenberg assumed responsibility for engineering and Wright and Harper 'undertook the certification of surveys and invoices and the work of inspecting and testing the whole of the plant before dispatch and after erection to ensure compliance with guarantees' (Klingenberg, 1916 [1913]: 195).

expected to push their total to 5000 million k.w.-hours, due to their acquisition of two additional mining companies (Modderfontain Gold Mining Company and the Bantjes Consolidated Mines) and new contracts for power supply to 'the Goldfields and Albu groups' (Klingenberg, 1916 [1913]: 195). In other words, Klingenberg's chain of reasoning for the establishment of the VFPC and the RMPS ends with the situation in which they appear on the scene just in time to meet the demand with a regional system of large central power stations.

Weinberger, on the other hand, claims that the idea to generate electric power at the Victoria Falls cannot be attributed to any single person or group. Instead, she attributes the gigantic regional power system on the Witwatersrand, and the strong role of the A.E.G. in its development, to the expansion of German imperialism, spurred by the close collaboration between German banks, government and industry (Weinberger, 1971). German banks were co-owners of mines in the Transvaal that, according to Weinberger, possessed one-sixth of the entire mining capital of the Transvaal. Deutsche and Dresdner Banks, together with Siemens and the A.E.G., respectively, had already established export organizations in the Transvaal by 1895.[113] In addition, Weinberger points out that the VFTPC and the RMPS removed the Siemens company as the main supplier of electrical manufacturing equipment in the Transvaal and established the monopoly of the A. E.G. The A.E.G.'s drive for capital expansion abroad was reinforced by the situation in Germany, where the construction of power stations was inhibited by the historically fragmented development of the electricity supply and legal restrictions (Weinberger, 1971: 62).

In Weinberger's account, the African Concessions Syndicate convened a board of experts to oversee the contracting process, and the American Ralph Mershon[114] and the German Georg Klingenberg independently proposed the best technical solutions. They reached an agreement to proceed with joint forces, and 'at the end of these negotiations was the establishment of the Victoria Falls Power Company in 1906' (Weinberger, 1971: 62).

---

113 Technical and Commercial Corporation Ltd., and United Engineering Company Ltd.
114 Hughes also mentions Ralph Mershon, as an engineer of Westinghouse (Hughes, 1983: 161) and graduate of the new electrical engineering programs (Hughes, 1983: 163).

With the Marquis of Winchester as Chairman, the head offices in London and the place of jurisdiction in Salisbury, Rhodesia, the VFPC appeared in public as a British company. Ten of the 12 members of the board of directors were British, and two were German. The German directors were Emil Rathenau and Hans Schuster. The African Concessions Company (on behalf of the BSAC) owned one million pounds of common shares, but the preference shares and debentures were possessed by German banks and the A.E.G. (Weinberger, 1971). A finance syndicate had been established to finance the VFPC, which in the beginning comprised 15 German and Swiss banks, headed by the Dresdner Bank (Weinberger, 1971: 63). In this way, the A.E.G. and the German banks owned the controlling interest and could dictate their conditions (Weinberger, 1971: 63).

Weinberger raises two important questions about this decisive point in the history of electrification in Johannesburg. Why would the the BSAC/ African Concessions Syndicate Company voluntarily concede rights and privileges to the A.E.G.? Why was the technical project commissioned to a German company despite the official preference for British companies, given that Ralph Mershon appeared to have submitted a technically equivalent solution? She mentions two factors that influenced this outcome: capital cross-ownership and the 1903 division of the world between the A.E.G. and General Electric.

One of the most powerful figures in the gold mining industry, Werner Beit, was financially involved in both the BSAC and the consortium of German banks that financed the VFCP. In February 1904, Union Elektrizitätsgesellschaft (UEG, founded 5 January 1892 as the overseas agency of the Thomson-Houston Electric Company (founded 1882)) merged with the A.E.G. in Berlin. All of the banks of the Union company were represented as directors of the A.E.G. At the time of the merger, Union Elektrizitätsgesellschaft was a subsidiary of the American General Electric Company. Rathenau had travelled to America in 1903 to negotiate with General Electric.

Weinberger writes that the first step that the VFPC took was to acquire the two existing power stations on the Rand, the RCEW and the GEPC. The particulars of this agreement are of interest, because they may be regarded as an excellent strategy by the VFPC to create a monopoly. Instead of paying for the power stations, the VFPC reimbursed its potential competitors by

involving them in the new enterprise; the RCEW was reimbursed with preference shares and debentures (175,000 and 175,000 respectively), and the
GEPC with 150,000 preference shares (Weinberger, 1971).

Renfrew Christie's history of electrification in South Africa dedicates an
entire chapter to the establishment of the VFTPC, entitled 'The Creation of a
Monopoly, 1905–1914'.[115] Christie emphasizes the close connections
between electrification, industrial development, labour and class in South
Africa. He considers the monopoly to be the logical result of the large capital
investment required for the establishment of an electric power industry.[116]
Accordingly, he presents the history of the VFTPC as resulting from the
struggle by different interest groups to obtain this monopoly. The winners of
this struggle, the A.E.G. and the German banks, in his view, had bought
'their way into the British colony of the Transvaal by paying the African
Concessions Ltd. an exorbitant price for the Falls Concession' (Christie,
1984: 31). Through this transaction, the BSAC 'was selling to the Germans a
British character for the power works' (Christie, 1984: 31). The VFPC project
to harness power from the Victoria Falls served to hide 'the real intentions' of
the company, which were 'to build a very large set of steam plants on the
Rand' (Christie, 1984: 31).

The Power Companies Commission and the Transvaal Power Act of 1910

In response to the VFPTC monopoly, the Transvaal government appointed a
Commission of Inquiry into the 'desirability of the establishment of large
electric power companies in the Transvaal', their effects on various industries

---

115 Christie's historical account is largely based on publications in the British journal
Electrical Review and correspondence between key players in this process: Professor Ayr-
ton, R. Hammond, Hermann Eckstein, Julius Wernher, Reyersbach, Harper, L. Phillips,
Winchester, Rathenau.

116 'Steam power was well-established on the Rand before developments in electrical
technology made steam less efficient than electricity [...]. The existence of steam made
the introduction of electric engines more difficult, because steam engines represented
large investments which had to be more than compensated for by increases in productivity, if electrification were to be profitable' (Christie, 1984: 13).

(gold, coal, railways and agriculture) and on labour (Power Companies Commission, 1910: 7). The Power Companies Commission also was expected to report on the state's powers and its involvement with such companies.[117] Christie (1984) has referred to the appointment of the Power Companies Commission and its findings as 'an exercise of accommodation' by the Transvaal government, 'whereby the new, and highly necessary, electricity monopoly was fitted into the existing relations of production, in such a way as to ensure the minimum of unnecessary disruption of vested capitalist interests' (Christie, 1984: 39). The Power Companies Commission was established by the governor general of the colonial government (Gentle, 2008: 55), 'his Majesty's High Commissioner for South Africa, and the Governor and Commander-in-Chief of the Transvaal', the Earl of Selborne (William Palmer) (Christie, 1984: 39).[118]

The Power Commission's report contained details on the authorized capital, assets, financial results and prospects of the VFTPC. It also recognized that 'It is obvious that the Victoria Falls Company intended, not only to buy up existing undertakings on the Rand, but also to increase and improve existing plant, to enlarge the sphere of operations, and to obtain as many as possible of the Rand mining companies as its customers' (Power

---

117   ,What powers and facilities (if any) should be conferred on such companies by the State; what restrictions and conditions (if any) should be imposed upon them, and what powers of expropriation (if any) should be reserved to the State' (Power Companies Commission, 1910: 7).

118   The Committee first heard 'the evidence of the proprietors of existing undertakings engaged in the supply of electric power and the promoters of proposed schemes, and after them all persons who might have objections against existing or proposed undertakings' (Power Companies Commission, 1910: 8). Written statements were also submitted on behalf of 'individuals and public bodies' (Power Companies Commission, 1910: 8). One of the individuals who provided evidence 'on behalf of' the Victoria Falls and Transvaal Power Company, in favour of the building of a monopoly system of large power stations was 'Professor' Georg Klingenberg. His evidence was referred to in the report. Charles Merz had also been approached for advice with regard to the work of the commission but he could not spare the time to travel to South Africa.

Companies Commission, 1910: 11).[119] The report also noted that 'the large holding of German firms is responsible for the fact that the supply of material for the new stations at Brakpan and Simmer Pan, as well as the contracts for erection, have fallen mainly to German contractors and manufacturers'. The commission was, however, satisfied 'that although the German subscribers of capital stipulated for German manufacturers and contractors, these latter would, in any case, have succeeded in open tender' (Power Companies Commission, 1910: 12). The Power Companies Commission pointed out that the articles of association of the VFTCP were 'of the widest character' (Power Companies Commission, 1910: 11) and those of its subsidiary, the Rand Mines Power Supply Company, were 'as various in character and as wide in scope as those of the parent company' (Power Companies Commission, 1910: 12).

Despite these reservations, the report emphasized the importance of cheap electric power for industrial development and concluded in favour of new technology,[120] centralized power generation and distribution,[121]

---

119 'The existing power companies have taken steps which have had the effect of eliminating competition within considerable areas, with the result that they are now in the possession of what is practically monopolistic power' (Power Companies Commission, 1910: 27).

120 'The greater simplicity and convenience of electric power is a technical advantage which has been adduced by several of the engineers who gave evidence' (Power Companies Commission, 1910: 16).

121 'The evidence before the Commission is unanimous that in comparison to the cost of an average mine where the size of generating plant does not exceed a few thousand kilowatts, the cost of producing power is greatly reduced when running plant ten to twenty times as large' (Power Companies Commission, 1910: 15), and 'Your commission are satisfied that the supply of cheap electric power to the mines will, on account of the economic and other advantages adduced, help to lead to an expansion of the mining industry' (Power Companies Commission, 1910: 16).

a monopoly of electricity supply on the Witwatersrand[122] and the need to stipulate government regulations.[123]

Following the report of the Power Companies Commission, the Transvaal Colonial Government passed the Transvaal Power Act of 28 May 1910.[124] It supported the construction of large electric power stations under the monopoly of the VFTPC and the A.E.G. under government supervision. Although the operational expansion of the VFTPC was allowed, the company was to be expropriated after 35 years. Under this Act, the generators and distributors of electricity in specific areas had to apply for a licence from the Power Undertaking Board.[125] Local authorities retained the exclusive right of supply within their boundaries. The Transvaal Power Act was passed only three days before the establishment of the Union of South Africa on 31 May 1910 (Christie, 1984).

---

[122] 'The principle should be adopted by the State that no legal monopoly be allowed within the area of supply, but at the same time, unless it is in the public interest, a competitor should not be allowed to enter the area of supply' (Power Companies Commission, 1910: 32). The report of the Power Companies Commission makes reference to the English Lighting Act of 1882, 1888, and 1899, the statutes of which 'specifically declare that no legal monopoly shall be granted to the undertaker' (Power Companies Commission, 1910: 29).

[123] 'Since the supply of electric power thus leads to the establishment of a virtual monopoly in a commodity which has become practically a necessity of modern civilization, it should, while being left as far as possible to private enterprise, at the same time be placed under Government control and subjected to regulations which shall secure the equitable supply of power, the public safety, and public interests generally' (Power Companies Commission, 1910: 26).

[124] 'This Act should constitute the requisite controlling and regulating authority under the direction of a Minister of the Crown, and should define its limits and the scope of its discretion in such a manner as to enable it to make general regulations and special provisions for the licensing, control, and supervision of power undertakings subject to such principles as may be laid down in the general Act in order to secure the public rights and interests' (Power Companies Commission, 1910: 30).

[125] 'It seems desirable to the Commission that every power company operating in this Colony should be registered in the Transvaal under limited liability laws of the Colony, and should have its head office in this Colony' (Power Companies Commission, 1910: 31).

The Act also regulated prices for consumers by introducing an auto-matic method of profit sharing as the basis for controlling the operations of power supply undertakings, which would continue to influence the econom-ics of the electricity supply in the decades to come. This method 'does not involve the continual investigation and litigation necessitated by the systems adopted in America and certain other countries' (Price, 1941: 272).

Technological empire and 'the native'

The VFTPC monopoly was formed in Johannesburg during the years when the Transvaal was still a British Colony. Companies of the two otherwise contending imperial nations, Great Britain and Germany, allied to establish this gigantic electric power scheme. The foreign monopoly was legally endorsed through the Power Companies Act just a few days before four sep-arate British colonies were joined to found the Union of South Africa as a dominion of the British Empire.

At the time of Union, Johannesburg was already racially segregated.[126] After Union, the different segregation policies of the four provinces[127] gave way to national segregation policies (Christopher, 1988), such as the Mines and Works Act of 1911 (sometimes referred to as the Colour Bar Act) and the Natives Land Act of 1913 (Act of Parliament of South Africa), the first major piece of segregation legislation passed by the Union Parliament.[128]

---

126 '[...] in Johannesburg [...] private compounds and shanty towns housed conside-rable numbers of Black and other persons. In 1902, the largest shanty town, "the Insani-tary Area", housing a mixed population, was demolished and a new all-White township, "Newtown", took its place. Blacks and Coloureds were moved to new locations set aside for them as singly race residential areas. [...] effectively removing a portion of the Black population from the sight of the White citizens' (Christopher, 1988: 155).
127 'Urban racial segregation in South Africa is complex and bound up with urban ori-gins and development. Prior to 1910 the four colonies which merged to form the Union had evolved different approaches to racial segregation ranging from passive socio-econo-mic segregation to legal enforcement' (Christopher, 1988: 151).
128 The Act decreed that only certain areas of the country could be owned by natives. It created a system of land tenure that deprived the majority of South Africa's inhabitants of the right to own land.

In 1910, the British Empire '[comprised] a fourth part of the habitable globe and almost a third of the numbers of mankind' (Lyttelton, 1911: 2), and 'the British dominions in South Africa, south of the Congo State, [extended] to some 1,200,000 square miles of country' (Hely-Hutchinson, 1911: 93). Birchenough, a member of the Board of Directors of the VFTPC considered 'the gold-bearing reefs and deposits of the Witwatersrand and of Southern Rhodesia' as 'among the wonders of the modern world' (Birchenough, 1911: 134). Newspaper articles on the Victoria Falls power project reveal the imperial disposition of this idea. An article in the Chicago Sunday Tribune of 28 November 1908 noted that the Victoria Falls amounted to five times Niagara's power capacity and '[...] may be used for the building up of an industrial empire which will affect the whole of this part of the continent' (Carpenter, 1908: 2).[129] The article further stated:

> The financial arrangements for harnessing the Niagara of the Zambesi have been completed. Within the last few months a London syndicate has been formed, with a paid up capital of 15,000,000 dollars, and surveys are now making for the installation of one of the greatest electrical plants of the world. The various power companies of the Rand have been purchased, and the new syndicate practically controls the power possibilities of South Africa (Carpenter, 1908: 2).

The New York Times in 1912 summarized the Victoria Falls power project as the largest undertaking outside of America:

> The great power scheme at Victoria Falls is not yet under way, and according to Mr. Hadley [Managing Director] it is not likely to be pushed until the country adjacent to the Falls gets other population than what it has at present. Just now the wild tribes don't need electricity, and lions, elephants, and giraffes cannot find any use for it, so the company [VFTPC] is devoting its attention mainly to making power in the Transvaal, and selling electricity and compressed air to the gold mines on the Rand (New York Times, 1912).

---

129   The newspaper article on *Harnessing the Victoria Falls. Industrial Revolution Likely to Follow the Utilisation of the Tremendous Force*, by Frank. G. Carpenter, published in the Chicago Sunday Tribune of 28 November 1908, reads: 'This power here, at the same rate, would daily equal the force of 1,000,000 tons of coal, to that, figuratively speaking, a million tons of black diamonds are dropping down into this gorge every twenty-four hours' (Carpenter, 1908: 2).

The original idea to 'harness' power from the Victoria Falls had formed part of Cecil Rhodes' vision of a British Empire from the Cape to Cairo.[130] This vision had led to the BSAC's annexation of Mashonaland and Matabeleland and its attempt to populate the country with white settlers. The railway reached the Victoria Falls in April of 1904, and the Victoria Falls Bridge across the Zambezi River was completed and opened by the President of the British scientific Association, George Darwin (son of Charles Darwin) only one year later, in July 1905.

The realization of the idea to build railways to, and a tourist resort at, the Victoria Falls at the beginning of the 20th century must be pictured. McGregor presents the idea of the railways reaching the Victoria Falls as provoking 'fantasies of economic growth, and global comparisons' (McGregor, 2003: 725):

> [...] the resort commemorated nineteenth century explorers [...] was significant in establishing a lineage stretching backwards, but it also reflected a new sense of distance from, and progress since, the age of the Victorian pioneers. The early twentieth century resort celebrated modernity, the achievements of colonial science and command over nature, epitomised by the railway, the bridge over the Zambezi [...] (McGregor, 2003: 725).

The local inhabitants of the area around the Victoria Falls, the Leya, had been removed from the lands that they had occupied around the waterfall: 'What is notable about the tourist images of scenery, leisure and luxury at the Victoria Falls, particularly on the South bank, is the exclusion of all Africans (other than the occasional glimpse of Lozi royalty or 'dusky servants'), and the failure to mention those who bore the costs and laboured to build the resort' (McGregor, 2003: 733).

Although the VFTPC ultimately did not exploit the Victoria Falls, the original idea continued to inspire the company, as is indicated by the retention of its original name and reputed purpose. The electrical industry and

---

130 'And in years to come, when industrial cities spring up north and south of the present township at the Falls, the fruits of the undertaking may be colossal. [...] it is not surprising that the scheme is one having a peculiar fascination for men of large prevision like the late Mr. Cecil Rhodes, and that engineers and other practical men have become enthusiastic since their interest was aroused' (Wills, 2005: 265).

technology established in Johannesburg was designed to meet the interests of settler populations. Although British and German colonial policies in Southern Africa pursued different strategies and ends, they shared a sense of social entitlement to impose colonial rule over the indigenous populations, which they referred to as 'the native' or 'the kaffir'.

The VFTPC's Simmerpan and Rosherville power stations were built in tandem with 'residences and quarters' for its employees (Hadley, 1913: 30).[131] At the time, 'public transport was limited to a horse-drawn bus to and from Cleveland, and stables were provided for employees who owned horses'.[132] The transport of materials was handled by 'a special department' that employed 'two motor lorries and 50 mules and horses' (Hadley, 1913: 31), and 'a fleet of 14 motor-cars [was] maintained in constant service for the use of those officers and engineers of the Company whose duties [necessitated] visiting the different parts of the system' (Hadley, 1913: 30–1). VFTPC employees were categorized into the following three groups.

1. A small, but highly skilled 'officers corps' consisting of engineers of whom the majority were British immigrants from the industrial areas of northern England and Scotland.
2. A middle class consisting mostly of local skilled white supervisors, clerks and artisans.
3. A large group of unskilled labourers.[133]

These employment categories essentially reflected the threefold racial classification of population groups at Union, introduced in the 1911 census.

---

131   'The Company realizes the importance of welfare work and its influence on the conditions of the life of the staff. They give a generous support to recreation and sport, and facilitate in every way the promotion of social intercourse among all classes of the employees. Some 60 residences and quarters have been built by the Company at the various power stations, and at each station a boarding-house and recreation-rooms are provided. Generally speaking, the conditions of life compare very favourably with those of an engineer on the mines' (Hadley, 1913: 30).

132   http://www.eskom.co.za/sites/heritage/Pages/VFP-%281%29.aspx., accessed 15.9.2015.

133   Quoted from Eskom's heritage site at http://www.eskom.co.za/sites/heritage/Pages/VFP-(1).aspx (accessed October 2015).

1. Europeans (Whites) – descendants of persons originating in Europe who were recognizably white.
2. Natives (Blacks) – descendants of persons indigenous to Africa, with the exception of the Khoisan peoples of the Western Cape, who were included in group 3.
3. Coloureds – other persons not included in the first two categories. The two main subgroupings, recognized in subsequent censuses, were Cape Coloured (the predominantly mixed population descended from Europeans, Africans and Asian slaves), and Asians, immigrants predominantly from India (Christopher, 1988: 151–2).

Accordingly, 'the towns and cities of South Africa were, in 1911, all of European foundation and constructed for the European colonists to live and work in' (Christopher, 1988: 152). Temporary quarters were established for the non-white society, and segregated residential areas were defined. At this time, Johannesburg had 'marked racial residential zoning restrictions which strongly influenced population distributions, as few Blacks or Indians were able to acquire property in areas where Whites sought consistently to exclude them' (Christopher, 1988: 156). Christopher considers mining settlements such as Johannesburg to have been much more segregated than other towns because of the 'presence of large numbers of Blacks housed in segregated company compounds' (Christopher, 1988: 165). The system of compound housing had been adopted from the diamond mines in Kimberley.

In 1907, the German colonial secretary, Richard Dernburg, designated technology as the most important ancillary science of the German colonizer.[134] He considered cultural progress to have affected a change in the colonial method: 'Whereas previously colonization took place by methods of destruction, today colonization takes place by methods of conservation, and these equally include the missionary, the doctor, railroads, machines, in

---

134 'Die Technik ist vielleicht die wichtigste Hilfswissenschaft des Kolonisators. Wir haben den Bohrtechniker und den Windmotor. [...] Wir haben den Elektrotechniker [...]' (Dernburg, 1907: 11).

other words, the advanced theoretical and applied sciences in all fields' (Dernburg, 1907: 9, my translation).[135]

In 1908, Bernhard Dernburg was accompanied on his tour of Southern Africa by Walther Rathenau, the son of Emil Rathenau (purchaser of Edison Patents for Germany and the founder of the A.E.G.), a member of the board of directors of the A.E.G., the head of the division on central power stations in 1899 prior to Klingenberg, a member of the governing board and later the board of directors of the Berliner Handelsgesellschaft and members of the board of directors of more than 100 companies, including the A.E.G., the industrial undertaking that received financing from the Berliner Handelsgesellschaft. Along with the Deutsche Bank, the Berliner Handelsgesellschaft had been involved in the syndicate that financed the establishment of A. Goerz GmbH in 1892, the company for which Siemens and Halske had built the first power station in Johannesburg.

On the occasion of their trip through Southern Africa, Rathenau was hosted by Louis Botha (who was to be the first prime minister of the Union of South Africa) and Jan Smuts (prime minister of South Africa in 1939). Rathenau and Dernburg were also received as private guests of Lionel Phillips (Christie, 1984: 43). Dernburg, Rathenau and Phillips shared a sense of entitlement of imperial nations over the indigenous populations of the Transvaal and Southern Africa, as indicated by the following quotes.

> Because the black man does not know investment capital, return on equity, administrative expenses, depreciation, settlement of working hours, his production cost is covered with his daily subsistence (Rathenau, Reflexionen, 2008: 149, my translation).[136]

---

135 '[...] ist es eine Freude, zu konstatieren, daß mit dem kulturellen Forschritt in der Welt auch die Kolonisationsmethoden eine große Wandlung haben durchmachen können. Hat man früher mit Zerstörungsmitteln kolonisiert, so kann man heute mit Erhaltungsmitteln kolonisieren, und dazu gehören ebenso der Missionar, wie der Arzt, die Eisenbahn, wie die Maschine, also die fortgeschrittene theoretische und angewandte Wissenschaft auf allen Gebieten' (Dernburg, 1907: 9).

136 'Denn der Schwarze kennt weder Anlagekapital noch Verzinsung, Verwaltungskosten, Abschreibungen, Zeitverrechnung. Seine Erzeugungskosten sind gedeckt, wenn er sich den Tag über ernährt hat' (Rathenau, Reflexionen 2008: 149).

But the native is the most important object of colonisation [...]. Because slavery has been abolished – thank God – and suitable labourers can only be obtained either from other colonies or one's own, and because manual power of the native is the most important asset, we have here an eminently important problem (Dernburg, 1907: 6–7, my translation).[137]

There are two respects in which even the uncontaminated Kaffir is low in the scale, according to our moral standards. He has no regard for the truth, and he is treacherous by nature. His smiling face is no index to his reflections (Phillips, 1905: 102).

According to the present standards of measurement, the natives are in all respects, in strength, in stamina, in capacity, and in acquirements, inferior beings, and, if that be acknowledged, the mission and the right of the white man lies in governing them with justice and firmness, in giving them a well-defined legal status, in promoting their intellectual and industrial training, in stimulating their power of self-control, and in inculcating the principles of morality. For the present they are separated from white men, not by colour, not by wealth, not even fundamentally by education, but by a gulf of profound mental dissimilarity (Phillips, 1905: 137).

Although there are no records that present the views of the A.E.G.'s designer of the gigantic power scheme in Johannesburg, Georg Klingenberg, on 'natives', he may be seen to follow suit with these previous quotes. His technological scheme did not consider 'the native' as a member of society but only as a labourer:

Notwithstanding the many impurities it contains, the cost of mining the coal is very small; the low depth at which it is found, cheap native labour, chiefly, however, the great thickness of seams, which make all constructional work unnecessary, keep the mining costs very low (Klingenberg, 1916 [1913]: 197).

---

137 'Nun ist aber der Eingeborene der wichtigste Gegenstand der Kolonisation [...]. Denn da die Sklaverei – Gott sei Dank – abgeschafft ist, die geeigneten Arbeiter also nur entweder auf dem Wege des Kontakts aus anderen Kolonien, oder aus der eignenen bezogen werden können, und die manuelle Leistung des Eingeborenen das wichtigste Aktivum bildet, so liegt hier ein eminent wichtiges Problem' (Dernburg, 1907: 6–7).

1914 to 1930: Enter Escom

In the development of every new country the first necessity is the establishment of means of communication – roads, railways, telegraphs. At a later stage, depending upon the labour, raw materials and markets available, manufactures spring up. In order to develop and flourish, manufactures nowadays need cheap power.
Merz and McLellan, 'Electric Power Supply in the Union of South Africa', 1920.

The First World War

The First World War brought South Africa 'higher productivity, new products, new markets, more machinery and greater use of energy, especially electricity'.[138] (Christie, 1984: 52). The war in Europe spurred the growth of local industry, and mining and agricultural production increased. After 1917, the gold mining companies undertook to 'increase mechanization of mining' and required more electric power (Christie, 1984: 56). Jackhammers were introduced, and working hours were prolonged. The VFTPC was directly influenced by these changing conditions: '[...] when gold boomed so would the VFTPC's fortunes. Equally, strikes would hit both' (Christie, 1984: 57). South Africa also developed new industries that required electricity during and after the war. Consumption increased because of 'a heavy industrial infrastructure [...] under state ownership or stimulus, in the fields of metallurgy, transport and energy' (Christie, 1984: 51).

These developments were accompanied by a series of strikes involving the gold mines, the coal mines and the VFTPC. In May 1915, employees of

---

138 'The wars brought state research into problems of local industry, they brought state co-ordination and planning of production, they brought fuller employment and higher effective demand. World financial disruption spurred on the creation of local capital. Competitive imports were excluded temporarily by shipping problems. The need for soldiers meant that machinery was used wherever possible in production, releasing men for military service. In all, the crises formed by the wars led to higher productivity, new products, new markets, more machinery, and greater use of energy, especially electricity' (Christie, 1984: 52).

the VFTPC went on strike 'demanding that the Company should cease to employ Germans', and in 1918, 'Johannesburg municipal power-workers struck for higher wages' (Christie, 1984: 55). In April 1919, the Johannesburg municipal power station workers went on strike. At the same time, a 'mass pass-law resistance campaign' was launched by the Transvaal Native Congress (Christie, 1984: 60). In early 1922, strikes were held by workers at collieries, gold mines and electric power stations of the VFTPC. Failed negotiations led to an armed uprising, known as the Rand Revolt. Power stations 'were an important focus of much of the fighting' in these events (Christie, 1984: 66). The revolt was put down forcefully by Prime Minister Jan Smuts. According to Christie, the strikes ironically 'moved the mine-owners further to control the labour-process by using machinery' (Christie, 1984: 67). This, in turn, increased the gold mines' demand for electric power from the VFTPC. After the strike and the Rand revolt in 1922, the VFTPC continued to grow steadily for 20 years (Christie, 1984).

After the war, the regulation of the electricity supply was considered at the national level for the first time, spurred by the issue of railway electrification. The Electricity Act of 1922 was passed, superseding the Transvaal Power Act of 1910. This new Act provided for the establishment of a national Electricity Commission (Escom) and an Electricity Control Board. Escom was authorized to sell electricity to consumers at neither profit nor loss. When the demand for electricity by the gold mines increased after 1922, Escom and the VFTPC built a new power station at Witbank. The equipment for this power station and the electrification of the railways was supplied by British manufacturers (Christie, 1984: 89).

The First World War ended the VFTPC's connection with the German manufacturing industry.[139] The war greatly impacted Germany's international industry, and 'prewar German interests in electric utilities [were] replaced by foreign capital intermediated by companies domiciled in the

---

139   In 1915, the managing director of the VFTPC would state, '"There is a popular idea that the VFTPC is a German company. This is entirely wrong. There is not a German Director on the Board, and it is only on account of the Debentures still held by the German banks that any connection with Germany still exists."' (Christie, 1984: 56, quoting Bernard Price, 1915).

homelands of the victors, in countries that had been neutral, or in countries that had been occupied by Germany [...]' (Hausman et al., 2008: 137). In August 1914, Britain forbade trade with Germany (Christie, 1984: 56). The German interests in the VFTPC 'passed to British ownership and control (with the control now linked with British South African mining interests)' (Hausman et al., 2008: 139).[140] Prior to the war, German investors 'had used or maintained British registration as they became owners of the equity of certain free standing companies in electric utilities' (Hausman et al., 2008: 139). The British government had promoted these processes as part of their attempt to foster the basic industries in the empire (Christie, 1984). In 1918, a British Departmental Committee on the Electrical Trades issued a report that considered the effect of the war on 'the British electricity industry's biggest competitor' (Christie, 1984: 52). The report regarded electricity as 'crucial to the industrialization of South Africa' (Christie, 1984: 54). Jan Smuts, a member of the Imperial War Cabinet, saw this report and 'took steps leading to the creation of Escom' (Christie, 1984: 54).

The Merz Reports on Railways (1919) and Electrification (1920)

In 1917, during the war, the British consulting company Merz and McLellan was appointed to report on the possibilities and advantages of electrification of the South African Railways. F. Lydall completed the report for Merz and McLellan in 1919. Merz discussed the report at several meetings in London with General Botha, General Smuts and Sir William Hoy, after the Versailles conference (Rowland, 1960: 87; Christie, 1984: 79). Merz subsequently travelled to South Africa for further conferences and was again commissioned by

---

140   At the imperial level, industrial research committees were established in England. The committee on electrical trades argued that the war would destroy the British electricity industry's biggest competitor, Germany, which, because of the relative backwardness of Britain and the financial power of Germany, had obtained British Empire contracts before the war, 'including the well-known case of the Victoria Falls Company' (Christie, 1984: 52).

General Smuts in August 1919 to undertake a study of the general supply of electricity in South Africa (Rowland, 1960: 88; Christie, 1984: 81).[141] The report was submitted in April 1920 and recommended that the South African government adopt a policy to establish 'a system of power supply [...] before the industrial development of the country [had] proceeded any further' (Merz & McLellan, 1920: 19). Merz considered South Africa to be in an advantageous position because it could enact legislation before industrial development and thus without interference of a variety of interests.

> In Great Britain and the older countries generally, the fact that electrical legislation has had to follow industrial development, instead of keeping pace with or, better still, preceding it, has seriously hampered the development of manufacturing. South Africa is fortunate in being able at the present times to lay down its power supply system almost at the foundation, instead of having to superimpose it upon an old-established industrial fabric (Merz & McLellan, 1920: 19).

To regulate and unify the electricity supply in South Africa, Merz recommended an Act of Parliament to establish 'a body similar to the Electricity Commissioners which had been set up in Britain under the Electricity (Supply) Act 1919' (Rowland, 1960: 88). Legislation was necessary to impose 'technical uniformity, limit prices, control the raising of capital and deal with water rights, wayleaves, and expropriation and the like' (Christie, 1984: 81).[142] The Merz report argued in favour of large power stations for

---

141 'The object of this Report is to consider the possibility and best means of organizing the supply of power to the industries both existing and likely to develop in the Union of South Africa' (Merz & McLellan, 1920: 1).

142 '[This Act] should deal with rights of way, compulsory acquisition of land, consents of local authorities, breaking up of streets, the use of water, statistical returns, powers to undertakers to co-operate and make agreements with other producers and industries; and their responsibility as regards issue of capital, rates and conditions of sully, tariffs, etc. It should authorise the fixing of standards of pressure and frequency' (Merz & McLellan, 1920: 17).

economic reasons.[143] Merz advocated the development of an electric power supply in South Africa as follows.

> The importance of developing the supply of electric power in the Union arises from the need for encouraging its existing manufactures and attracting new and permanent industries as rapidly as possible. This latter has been emphasized by yourself, and by many important public bodies, who have stated that *in no other way can an adequate white population be obtained*. It is therefore a policy which justifies the expenditure of money in advance of to-day's actual needs (Merz & McLellan, 1920: iii; my emphasis).

In this way, Merz linked the necessity of large-scale, strategic electrification to the development of the 'white population' of South Africa. In the report, he refers to the Electricity (Supply) Act that had just been passed in Britain, which he considered to be 'not without its bearing on South Africa, in view of the similarity of the general principles involved' (Merz & McLellan, 1920: 18).[144] The report regularly refers to South Africa as a 'new country' and contrasts it with 'Europe' (Merz & McLellan, 1920: 15) or 'Great Britain and the older countries generally' (Merz & McLellan, 1920: 19). Merz repeatedly emphasizes the advantages of the South African situation compared with Great Britain.

> This being the existing state of affairs, it may be said that the Union is fortunate because in many ways it is in a much better position than Great Britain and other old countries as regards the possibilities of initiating a new policy of electrical supply. In these older countries vested interests have been created to a much greater extent, and must be considered to a greater degree than in South Africa. In Great Britain the industry is still to a large extent governed by legislation passed over 30

---

143   'The Report shows that in order to obtain electric power cheaply its production must be, as far as possible, concentrated in large plants, and the distribution network must be extended as widely as possible' (Merz & McLellan, 1920: iii).

144   'In this connection the recent legislation in Great Britain is not without its bearing on South Africa, in view of the similarity of the general principles involved. Under the British "Electricity (Supply) Act, 1919," a Board of Electricity Commissioners is appointed, to whom have been transferred all the powers relating to electric supply hitherto exercised by the Board of Trade and other Government departments' (Merz & McLellan, 1920: 18).

years ago, which was framed in a parochial spirit, and has much retarded progress. The reorganization of electric supply was only undertaken by the British Parliament last year, and in view of the considerable electrical development which has already taken place, must be a slower and more difficult process than in South Africa. In the latter country electrical and industrial development is fortunately at an early stage, and advantage can be taken of the experience of Great Britain and other countries to void their mistakes, and of the conditions in South Africa to exploit its advantages (Merz & McLellan, 1920: 12).

The report recommends standardizing a 'comprehensive system of power supply' across the Union of South Africa before industrializing the country, including such matters as the type of current and frequency, and the standard voltages for transmission and distribution (Merz & McLellan, 1920: 13).[145]

The Electricity Act of 1922, Escom, the Electricity Control Board, and Witbank Power Station

A committee under the Chairmanship of Sir Robert Kotze, Government Mining Engineer, drafted and submitted an Electricity Bill to the Minister of Mines and Industry on 4 March 1921. After various negotiations, the amended Bill was presented to Parliament in 1922. The Merz report recommendations 'were put into effect by the passing of the Electricity Act No. 42 of 1922', which became effective on 1st September 1922 (Jacobs, 1941: 261).

The Act provided for the creation of the Electricity Supply Commission (Escom) and an Electricity Control Board. The functions of Escom were: 'The establishment, maintenance, and working of undertakings for the supply of electricity to the Government and any other bodies or persons whatsoever in the Union of South Africa, and the investigation of new or additional facilities for the supply of electricity and for the co-ordination or co-operation of existing undertakings so as to stimulate the provision of a cheap and abundant supply of electricity' (Jacobs, 1941: 261). Escom 'was obliged to establish and work electricity undertakings to supply railways, government

---

145 Merz points out the problems encountered in Great Britain where 'numbers of systems are to be found employing frequencies varying from 25 cycles to 133 cycles' (Merz & McLellan, 1920: 13).

departments, local authorities, industrial undertakings and private persons' (Christie, 1984: 84). It would be responsible for the evaluation of, and reporting on, tenders for electricity undertakings to the Electricity Control Board (ECB). The ECB[146] was responsible for the control and licencing of electricity undertakings and for the examination of tariffs (Jacobs, 1941).[147] Decisions on these tenders were to be made by the Governor-General. Escom was obliged to operate at neither profit nor loss. The Act afforded municipalities the right to provide electricity in municipal areas, on condition of approval by the provincial administrator (Christie, 1984).

Witbank Power Station

During this same period, the VFTPC anticipated an increasing demand for electricity and started planning a large new power station at the Witbank coalfield. By the time the Electricity Act came into force, the necessary 'agreements for the supply of coal, water and other requisites had been concluded, designs had been prepared and everything was in readiness to proceed' (Price, 1941: 272). In 1923, after the 1922 strike had been put down, the VFTPC applied to the Electricity Control Board for authorization to build the power station. The Electricity Supply Commission was hurriedly appointed, with Hendrik van der Bijl as chairman. According to Bernard Price, general manager of the VFTPC, one of its first actions was 'to hold up the hearing of the Company's application and to engage Messrs. Merz and McLellan as its Consulting Engineers' (Price, 1941: 273). With the assistance of Merz, who travelled to Johannesburg for this purpose to advise Escom, 'the heads of an agreement were negotiated which laid the foundation for the co-operative arrangement under which the Commission and the power Company have been operating' (Price, 1941: 273).

---

146   The first members of the ECB were appointed soon after the Act had been passed and included Sir Robert Kotze, H.J- van der Bijl, J.A. Vaughan and A.C. Marsh (Christie, 1984: 84).

147   'When a licence was applied for, the ECB would hold a public hearing of objectors and arbitrate between rival applications. The ECB would fix standard prices and enforce the sharing of surplus profits with consumers' (Christie, 1984: 84).

The agreements between the VFTPC and Escom for the Witbank power station were 'of far-reaching significance in the development of power supply on the Rand' (Price, 1941: 273). The negotiations mediated by Merz also involved the gold mines, collieries, the Johannesburg municipality, and the South African Railways (Christie, 1984: 87), and they resulted in 'a carefully drafted synergy between Escom and the VFTPC' (Gentle, 2008: 58). The Witbank power station would be financed and owned by Escom, but it was to be built, operated and maintained by the VFTCP on behalf of Escom (Christie, 1984: 87; Price, 1941: 273).[148] Electricity from the Witbank power station would be supplied by Escom to the South African Railways, to collieries, and to industries and other consumers in Witbank. The municipality of Johannesburg would also be able to purchase power from Escom when required. The VFTPC was obliged to purchase from Escom and to transmit all remaining power to the gold mines (Christie, 1984: 87). Under this arrangement, Escom assumed the role of the provider of electricity for the collieries and the railways, and the VFTPC was the supplier of power to the gold mines (Gentle, 2008: 58). The licence to establish and supply electricity from the Witbank station was granted to Escom. Contracts for the Witbank Power Station were given to British manufacturers and construction companies.[149]

---

[148] 'An equitable basis was worked out for dividing the cost of producing the energy at the Station, including capital charges, between the parties according to their respective use of the plant, so that the overall result was very nearly the same as it would have been had the Power Company invested its own capital and incurred capital charges thereon at the rates allowed to it when arriving at surplus profits for division with its consumers' (Price, 1941: 273).

[149] These included Babcock & Wilcox, Ltd. (boilers and boiler house accessories, pipework and pumps, steel buildings and coal and ash handling plants), CA Parsons & Co., Ltd. (steam turbines, generators, condensing plant and water cooling plant), Metropolitan-Vickers Electrical Co., Ltd. (transformers and power station switchgear and accessories), Alexander Jack & Co., Ltd. (overhead travelling crane), Drysdale & Co., Ltd. (vertical spindles pumps and motors), J. Blakeborough & Sons, Ltd. (valves, penstocks and fittings for pump house), Paterson Engineering Co. (filtration plant for domestic water), Stewarts & Lloyds, Ltd. (water piping, valves, etc.), British Mannesmann Tube Co., Ltd. (poles for overhead transmission lines from generating station to pump house), The Baughan Crane Co. (crane for pump house), A. Stuart, Germiston (construction of dam,

These principles of 'co-operative arrangement' were applied to subsequent new power stations on the Witwatersrand. Escom also established
power stations in Natal and Cape Town after 1924. Electricity supply in
Johannesburg would remain dominated by the VFTPC until 1948, 'essentially a foreign company making abnormally high profits [...]' (Gentle, 2008:
58) from its business in South Africa. Escom operated 'as freely as does any
private company' and was not 'in any sense a branch of the civil service'
(Jacobs, 1941: 261). Its only stipulations were the appointment of the Commissioners by the Governor-General and his approval of any loans taken up
by Escom for its undertakings. Essentially, Escom operated independently of
government, but it had the status of 'a public utility corporation initiated by
the Government' (Jacobs, 1941: 261).

The municipality of Johannesburg continued to generate and distribute
its own electricity, because '[...] neither Escom nor the VFTCP reached an
agreement for cooperative working with the Johannesburg Municipality [...]'
(Christie, 1984: 88). In 1923, shortly after the Electricity Act was passed,
Johannesburg municipality was given a licence to build a new power station
on Jeppe Street adjacent to the existing power station. The first electricity
from the 10 MW turbo generator at Jeppe Street Power Station was generated in September of 1927, and two further generators were completed
three years later. This independent network was only connected to the
VFPTC distribution system in 1934 (Christie, 1984: 89).

In 1928, another major consumer of electricity was established by the
South African state, the Iron and Steel Corporation (ISCOR). This state-
owned enterprise was chaired by Escom's chairman, Hendrik van der Bijl,
who subsequently 'co-ordinated the work of his two state-owned enterprises
in industrializing South Africa' (Christie, 1984: 93). Escom, ISCOR and the
South African Railways represented three major industries that were 'either
state-owned or had a major state-owned component' (Christie, 1984: 95).
ISCOR built its own power station, designed by A.M. Jacobs, an Escom

---

pump house, residential buildings, excavations, drainage system, cooling ponds and service reservoir and laying of pipeline), A. Bradbury & Co. (Pietermaritzburg, construction of
railway siding from Witbank Station to power station). (http://www.eskom.co.za/sites/heri
tage/Pages/Witbank.aspx), accessed 20.6.2015.

engineer (Christie, 1984: 93). By 1930, the total generating capacity in South Africa had risen from 162 MW (supplied by the VFTPC) to 418 MW (supplied by the VFTPC and Escom) (Troost & Norman, 1969) (see Table 6).

	1915 generating capacity (MW)	1930 generating capacity (MW)
Brakpan	31	52
Simmerpan	40	40
Rosherville	48	60
Vereeniging	43	51
Witbank		101 powered by duff coal, Escom owner, VFTPC constructor and operator; transmission line to Brakpan, where it linked up with the VFTPC distribution system
Colenso		60 (Railways Natal 1921–26) Built by SAR sold to Escom in 1927
Congella		24 (Durban 1928)
Salt River		30 (Cape Town 1928)
Total	162 MW (VFTPC)	418 MW (VFTPC and Escom)

Table 6: The total electricity generating capacity in South Africa in 1915 (supplied by the VFTPC) and in 1930 (supplied by the VFTPC and Escom) (Troost & Norman, 1969).

In 1938, Escom decided to build the Vaal power station, and in 1940, the Klip power station was completed. In 1948, Escom purchased the VFTPC for £13,500,000 (Gentle, 2008: 59), with the assistance of a (£15,000,000) loan granted primarily by the Anglo American Corporation. The Transvaal Power Act of 1910 had given the state the opportunity to purchase the private utilities by 1950. It was not until the late 1960s that the power stations started to interlock, and a national grid was achieved in 1973. During the 1970s, Escom 'undertook a major programme of expansion, building very large pithead power stations on the coal fields of the Eastern Transvaal and Northern Natal' (Gentle, 2008: 62). In the early 1990s, only 35% of South African households had access to electricity, and the new government embarked on a national electrification programme. Between 1994 and 1999, approximately 2.8 million households were connected to the national electricity grid. The level of electrification was increased to about 66% of households in 1999, and to 86% in 2014.[150]

"National" legislation

It has been remarked that in the Union of South Africa the native areas can generally be distinguished on a map by the fact that these areas are generally excluded from the otherwise well-developed railway system. The same is generally true of the electricity system in both rural and urban areas. Lydall's report on railway electrifications perpetuated the pattern [...]. Black peasants were accordingly denied the benefits of railways and power systems
Renfrew Christie, 1984.

During this phase, national legislation was introduced that would determine the conditions for electricity supply for the next 50 years. The VFTPC played a formative role in the establishment of the two government bodies appointed to regulate the industry. Had no foreign-owned electricity supply monopoly existed in Johannesburg, it is likely that national legislation would have taken another form. Just like Klingenberg, Merz advocated large power stations with a monopoly of supply (Christie, 1984: 81). Christie considers Merz and McLellan's recommendations to have '[...] in fact formed the basis for the non-Witwatersrand spatial expansion of Escom until 1945' (Marquard, 2006: 147), and '[...] laid the basis on which the provincial and national grids were built up' (Christie, 1984: 81). However, the influence exerted on this legislation by the British consultant Charles Merz was not confined to the technological expertise of his engineering company; he also represented British financial and industrial interests more broadly.[151] Other

---

150 Statistics South Africa: General Household Survey 2014.

151 'Merz was doyen of the British electrical engineers, and had long been mentor to Bernard Price. Merz exported British technology to the world, acting as consultant to many governments, municipalities and railways systems. His connections with the financiers of the City of London were perhaps more subtle than those of Rathenau's A.E.G. with the Deutsche and Dresdner banks, but they were no less close. Merz represented British finance capital, in which banks and manufacturers combined to export to the

interests represented in this process were the manufacturing industry, the Natal coal industry, De Beers, and various state officials, such as General Smuts, William Hoy, Hendrik van der Bijl and Robert Kotze (Christie, 1984).[152]

Merz had personal connections to the VFTPC through his former employee, Bernard Price, the general manager of the VFTPC.[153]

> Several reasons led to Merz's decision to open an office in London [...]. There was however another reason why the London office was regarded as important. It was that the partners knew that electricity for power was bound to be extended overseas as well as in this country, and if British electrical engineers were to justify their existence they would have to work for overseas clients as well as those at home. There were great areas in Australia, in South Africa and in Canada, where industrial development was slow, partly because a cheap and abundant source of power was lacking. If electrical energy could be made available, the development of the Empire, now known as the Commonwealth, would progress by leaps and bounds. Any overseas development could most conveniently be discussed in London (Rowland, 1960: 51–2).

In addition to being a consultant to the South African government, Merz was actively involved in electrification at a global scale:

> Merz was already interested in what was happening overseas. He had made an early visit to the USA in 1901 and later in 1906 revisited Chicago, where, at the request of

---

formal and informal empire, be it in Australia or Argentina, Eskom is as much the creation of Merz wishing to sell British technology as it is the creation of South African industrialists' (Christie, 1984: 76).

152 General Smuts was Prime Minister of South Africa, William Hoy was General Manager of the South African Railways, Hendrik van der Bijl would be Chairman of Escom from 1923, and Robert Kotze was Government Mining Engineer.

153 'I look upon Dr. Merz as one of the biggest men of his time and, in a special sense, the father of centralised power supply in the British Commonwealth. He was much more than an able engineer. He was a man of outstanding vision and balanced judgement with an insight into the essence of a problem which amounted almost to an instinct. [...] Throughout the world, in America, Australia, India, South Africa as well as in Great Britain he has left an indelible mark on power supply development [...]' (Price, 1941 in Jacobs, 1941: 271).

Samuel Insull, he prepared a comprehensive report on the electric system there. Insull had suggested that it would be useful to deal with ways of laying out a network of distribution lines so as to ensure continuity of supply [...]. In Ottawa he met too, Earl Grey's son-in-law, W. Grenfeld, whom he had known in England; and was asked to assist in some complicated negotiations on power supply to the gold mines on the Rand in South Africa. These jobs had been interesting. They had given Merz a new insight into many problems (Rowland, 1960: 52).

The first 'national' legislation of South Africa was shaped by a British consultant who viewed South Africa as a 'new' country that required electricity for industrial and social development.

South Africa possesses most of the requirements needed for the development of manufactures. It has wonderful natural resources, a splendid climate, and both internal and external markets for its goods; and although at present its white population is not proportionate to its area and possibilities, it is probably sufficient for its immediate manufacturing needs if power be more widely used, while it obtains valuable supplies of good native labour (Merz & McLellan, 1920: 1).

Following the promulgation of the new legislation, Merz and McLellan expanded their activities in South Africa and Rhodesia, and established offices in Johannesburg and Salisbury.[154]

The British consultant Charles Merz substantially influenced the subsequent history of electrification in South Africa. He prepared the ground for the establishment of Escom, the major institutional product of the first national legislation. Escom played a key role in supplying the gold mines with cheap energy in the years before the Second World War (Gentle, 2008: 59). It also set up the conditions that would give the municipalities a minor role and little influence in electrification during these years. The grand power scheme of Johannesburg was established on the Witwatersrand, the wider Johannesburg metropolitan area, without considering the requirements of

154   'As a pleasant contrast, however, the work in South Africa and also at Salisbury in Southern Rhodesia tended to extend, the war having given a great impetus to industrial development in South Africa. [...] The work in Rhodesia was originally dealt with from Johannesburg but as it grew the decision was made to establish a further local firm – Merz and McLellan (Rhodesia) with the same partners as the South African firm' (Rowland, 1960: 107).

urban or municipal Johannesburg. The appointment of Hendrik van der Bijl as chairman of Escom also marked the beginnings of a closely knit association of large industries under state involvement.

Escom's status and sphere of power as a government-instituted commission are significant. Escom was instructed to plan the development of electricity schemes at the national level, but it was classified as a private producer (Marquard, 2006: 132). The Electricity Control Board was mandated to licence electricity undertakings, but it ended up approving tariff increases for the VFTPC and Escom's tariff structure (Marquard, 2006: 133): 'In reality, the Board had almost no capacity (having 3 to 5 board members and one staff member) to undertake sophisticated economic regulation [...] it seems that the Board's time was taken up playing a mediating and facilitative role in resolving disputes concerning rights to supply, holding hearings on land expropriation for electricity infrastructure, and addressing consumer grievances' (Marquard, 2006: 133).[155] In effect, the new regulation entrenched the (foreign) private sector generation monopoly and expanded it with various 'undertakings' of Escom to sell electricity to mines, railways and municipalities. Escom possessed a high degree of autonomy,[156] it was not accountable to political institutions and processes, and its leadership had close ties with the government. One of the likely consequences of the informal and self-governing status of Escom was its decision to suspend development of a national grid until the early 1970s. Until 1950, electricity demand was dominated by the energy-intensive mining industries (Marquard, 2006: 138).

The association between the VFTPC and Escom ultimately put electrification in the service of the industrial development of the British dominion of the Union of South Africa and its white population of the British Commonwealth.

---

155   The ECB only enlarged after 1977 after a Board of Trade and Industries report into electricity tariffs, but even then 'the ECB was not successful in exerting [...] influence over Escom' (Marquard, 2006: 133).

156   'Oversight of the organisation was carried out by a 'Commission' of around five people appointed by the Governor-General (later the relevant Minister); this process of appointment constituted the only direct involvement by the government in Escom until the 1980s' (Marquard, 2006: 150).

3.3 "How Do Technological Systems Evolve?"

To answer the question 'how do technological systems evolve?', Hughes presented a systems history of electrification based on the case studies of Chicago, Berlin and London from 1880 to 1930. He used these case studies to develop his framework for studying technological change. This framework is based on a number of *themes* and *subthemes* that shape the history of electrification: 'Embodied in the different power systems of the world is a complex variation on major themes that keeps the technology from becoming homogeneous and dull and that provides the historian with the challenging task of description and interpretation' (Hughes, 1983: 17). He described the subthemes as 'in most instances related to questions often asked about technological systems and about the history of technology in general' (Hughes, 1983: 461).

However, Hughes does not discuss the difference between themes and subthemes, nor does he specify them for his study. The five phases of his system model might be viewed as major themes: invention and development, technology transfer, system growth, technological momentum, and maturity. Hughes also introduces the expressions reverse salient, critical problems, the culture of regional systems, the style of evolving systems, the system goal and the system builder to present his systems framework. Are these themes or subthemes that emerge from the study of electrification, or simply new concepts introduced by Hughes to analyse technological change? This question cannot be answered conclusively. Hughes' inexplicit use of terminology has methodological consequences for the analysis of additional case studies. It complicates the task of comparing Hughes' themes with themes that emerge from histories of electrification in other metropolitan areas. This, in turn, makes it difficult to improve Hughes' model or develop new concepts to analyse technological change.

Trouillot's analytical framework of geographies of management and imagination has yielded a number of recurring themes in the history of electrification in Johannesburg: gold mining, empire, national competition, companies, concessions and the 'native'. These themes will be discussed separately below, across the various phases of the history of electrification in Johannesburg. Then, these themes will be related to Hughes' model to develop concepts that are able to capture the dynamics, procedures and underlying forces of electrification in Johannesburg. Based on these considerations, alternative analytical and descriptive tools for the empirical analysis of technological change are proposed.

Hughes claimed that the 'change in configuration of electric power systems 1880–1930' constituted the 'formative years of the history of electric supply systems' (Hughes, 1983: 1). One of his brilliant insights was to assume that the history of electric power systems 'extends beyond national borders' (Hughes, 1983: x). Thus, his book addressed the question: 'How did small lighting systems of the 1880s evolve into the regional power systems of the 1920s?' (Hughes, 1983: 2). The objective of his book, therefore, was to explain the change in configuration of electric power systems during the half-century between 1880 and 1930 (Hughes, 1983: 2) by investigating major themes in the history of electrification of Chicago, Berlin and London. Hughes considered these themes to be relevant 'to history focused on technology more generally' (Hughes, 1983: 461) and applied his 'model of system formation and growth' as a 'mode of organization' to 'co-ordinate' them. Following Hughes' chain of argument, the following paragraphs identify the themes that help 'to explain the change in configuration of electric power systems during the half-century between 1880 and 1930' (Hughes, 1983: 2) in Johannesburg.

Themes in the History of Electrification of Johannesburg

Six recurring themes emerge from the history of electrification in Johannesburg: gold mining, empire, national competition, companies, concessions, and the 'native'. These themes are diverse and do not suggest any obvious connections. Neither do they appear to respond to Hughes' question concerning how the small lighting systems of the 1880s evolved into regional power systems. To claim that electrification in Johannesburg between 1886 and 1930 was driven by gold mining, empire, national competition, companies, concessions, and the 'native' does not make sense. Nor do these themes, prima facie, appear to be significant to the history of electrification in Chicago, Berlin or London. What, then, do these themes signify?

To investigate this question, the themes are assessed individually in more detail across the various phases of the history of electrification in

Johannesburg. Then, these driving forces in the history of electrification in Johannesburg are weighed against Hughes' themes and framework for studying technological change. At several key moments, the history of electrification in Johannesburg unambiguously and directly connects to the history of electrification in Chicago, Berlin and London. To make visible the themes identified in the history of electrification in Johannesburg, Hughes' framework needs to be supplemented by new concepts. These new concepts should provide new historical markers to the question of how small lighting systems of the 1890s evolved into regional power systems outside of the confines of the category of 'Western society'.

For the moment, the themes – gold mining, empire, national competition, companies, concessions, and the 'native' – appear to be rather remote from Hughes' topic of technological change. This has to do with the different frames of analysis employed in his histories of Chicago, Berlin and London on the one hand and the history of Johannesburg on the other hand. Hughes presents a systems history of regional power systems at three sites by assuming a 'seamless web' of technology and society. Hughes' 'rationale' for studying electric power systems is the assumption that the history of all large-scale technology (not just power systems) can be studied effectively as a history of technological systems. The rationale for studying the electrification of Johannesburg was the assumption that all large-scale technology (not just power systems) can be studied effectively as a history of the North Atlantic universal 'technology'.

Hughes' systems are connected in a network of interacting components, with defining system properties such as centralized system control, a common system goal, and system environment. His history claims to go beyond a mere history of connected internal and external forces, dynamics, growth and factors; it depicts 'a history of technology and society', shaped by the specific cultures of the societies under examination (Hughes, 1983: 2).

Instead of Hughes' setting of the network, or the connected system of components, the history of electrification in Johannesburg used the geographical scenery as backdrop. Following Trouillot, this study traces the history of electrification in Johannesburg through the intertwined lenses of the geographies of management and imagination. Electrification in this framework appears as a deployment of the North Atlantic universal 'technology'. Trouillot's geographies or lenses serve to trace North Atlantic universals, which are hard to conceptualize but seductive to use because they have the power to silence their own history. The challenge, therefore, is to 'unearth those silences, the conceptual and theoretical missing links that make them so attractive' and to reveal their 'hidden faces' (Trouillot, 2003: 36).

Hughes' notion of technology, as employed in his history of electrification and model for technological change, classifies as a *North Atlantic universal*. However, *North Atlantic universals* offer neither descriptive nor referential accounts of the world; instead, they offer visions of the world. These visions are prescriptive because they project 'an historically limited experience on the world stage' and thereby imply 'a correct state of affairs – what is good, what is just, what is sublime or desirable' (Trouillot, 2003: 35). Whereas Hughes' framework pursues answers to the question of how technological systems (particularly electric power systems) evolve, the framework proposed here in the analytical style of Trouillot seeks to ask how technology – or its deployment in the idea of electrification – sets the terms of the debate and restricts the range of possible responses.

The geographies of management and imagination might appear to be more abstract than Hughes' network or system of components. Indeed, the forthright, nuts-and-bolts image evoked by Hughes' terms for the specific case of electric power technology may have contributed to the lasting authority of his model. In addition, Trouillot's notion of the geography of imagination refers to an indefinite and therefore contested realm. Moving in this realm requires considerable abstraction. The abstraction, however, is not artificial or limitless; it mirrors very specific empirical circumstances. For example, consider the late 19th century idea to conquer land, to generate electric power at a place named the 'Victoria Falls' in Southern Africa (estimated by engineers to carry five times the horse power of the Niagara Falls) and to carry electrical current to gold mines over a distance of 600 miles. This idea appears both grandiose and abstract, but it helped to shape the history of electrification in South Africa (see Figures 5 and 6). Or, consider the idea to build a German electric power station in the late 1890s for German-owned gold mines in Johannesburg. The railway line from Cape Town to Johannesburg had only just been opened, and ox-wagons still provided the principal means of transport. This idea, too, may be considered to be abstract retrospectively, but it helped to shape the history of electrification in South Africa.

Figure 6: The Victoria Falls at the Zambezi River.

THE VICTORIA FALLS COMPARED AS TO LENGTH WITH OXFORD STREET, AND AS TO HEIGHT
WITH ST. PAUL'S CATHEDRAL.

Figure 7: The Victoria Falls at the Zambezi River compared to Oxford Street
and St. Paul's Cathedral.

The geographic scenery forces us to juggle different lenses to map various
scales, places and (at times imaginary) spaces from empirical material. This
procedure does not lead to a never-ending confusion of possible connections.
On the contrary, the case of electrification in Johannesburg has shown that
certain developments can only be explained by consulting the abstract realm
of the imaginary. For the case of Johannesburg, the specific nature of this

imaginary is most obviously exposed in the historical figures and institutions that influenced Hughes' history of electrification but also appear in the history of electrification in Johannesburg, such as Emil Rathenau, Georg Klingenberg and Charles Merz. Thus, the analytical step into the geography of imagination does not lead to unlimited possibilities in historiography, because this step maps levels of abstraction that are intrinsic to all empirical material. In fact, it is precisely this focus that allows us to shed light on circumstances that are otherwise overlooked.

Gold mining

Of all industries to which it is applicable, gold-mining is the one which has come mostly to my notice as wanting a continuous supply of power day and night, and often without any economical means of getting it except by electric transmission. In these cases it will often be profitable to the gold miner to pay a high price for his power.
George Forbes, 1898.
British consultant to the Niagara Falls power project and consultant to Cecil Rhodes on the Victoria Falls power project.

The first electric lights, the first power stations and the first regional power system in Johannesburg were all set up to serve the gold mines of the Witwatersrand. The evolution of small lighting systems in Johannesburg into a regional power system followed the dictates of the expanding gold mining industry. Johannesburg's *raison d'être* was gold. It was founded to accommodate the settler and migrant communities pouring into the Witwatersrand after the discovery of the gold fields in 1886. In 1914, the VFTPC and its subsidiary, the Rand Power Supply Company, sold 99% of its generated electricity to the gold mines. Christie aptly described this close relationship between gold and electricity during the First World War: 'If Britain needed gold, the South African gold-mines needed the VFTPC' (Christie, 1984: 58). In other words, the demand for gold spurred the demand for electric power. The gold mines were the principal purchasers for the electrification projects in Johannesburg between 1886 and 1930.

Gold in itself, however, was not a primary driver for electrification. The initiatives to build the first central power stations in Johannesburg did not spring from local requests for electric power. They originated from German and British imperial objectives in the Transvaal. The first regional power scheme in Johannesburg also did not grow from local power requirements. On the contrary, this power scheme resulted from a clever merger between several ideas that served to further British and German colonial interests in the Transvaal. Nevertheless, electrification in Johannesburg was presented in terms of the power needs of the gold mines. In this way, 'gold' set the terms of the debate for electrification in Johannesburg and restricted the range of possible paths to be taken. What afforded this mineral such discursive power?

The timing of the gold discoveries on the Witwatersrand was significant. Gold was discovered in the Transvaal at a time when the international economy 'was suffering from severe structural crises and rivalries among imperialist powers' (Van Helten, 1982: 530). During the course of the 19th century, Great Britain, the world's most powerful nation, had adopted gold as the monetary basis for its pound sterling, and other powerful nations, such as France and Germany, had followed suit (Van Helten, 1982). Gold came to dominate global monetary relations after 1870 and 'bolstered the Bank of England's hegemonic position in the international financial system' (Van Helten, 1982: 529). At the end of the 19th century, gold was not simply an end in itself; it had 'emerged as the basis of international payments among the leading industrial nations of the world' (Van Helten, 1982: 533). Gold became the lifeblood of the political economy of imperial nations. For this reason, the gold discoveries on the Witwatersrand propelled this previously unknown part of the world 'onto the centre stage' of the international economy (Van Helten, 1982: 530).

This centre stage attracted settlers, companies, investors, capital and manufacturers from Europe to exploit the profits of gold mining, and it imposed new social formations and hierarchies. Local and migrant African workers were employed in the gold fields under conditions that deprived them of rights.[157] Only twelve years after gold was discovered on the

---

157   'Tens of thousands of diggers and fortune seekers made their way to the Transvaal from the four corners of the earth and large numbers of French rentiers, British investors and German banks poured millions of pounds into a myriad of honest and dishonest

Witwatersrand, a remote region in Southern Africa that had to be reached by ox-wagon before railways arrived in 1892, the Transvaal 'was producing over one quarter of the world's annual output of gold' (Van Helten, 1982: 529). This project of enormous historical scale was orchestrated by South African gold mining companies, which branched out into a 'complex London network of brokerage, insurance, refining and marketing facilities of gold' (Van Helten, 1982: 548). The early history of electrification in Johannesburg was determined by the specific companies and institutions involved in this gold mining industry, such as Consolidated Gold Fields, Werner, Beit and Company, Eckstein, A. Goerz Co., the Rand Mines Co., the Central Mining and Investment Company, the South African Chamber of Mines, the British South Africa Company, the Deutsche Bank and the African Concessions Syndicate.

Hughes' history of electrification in Western society does not make reference to gold. On closer inspection, however, some interesting connections emerge. Thomas Edison developed businesses in the field of mining technology and ore-crushing processes.[158] Furthermore, gold is not mentioned in Hughes' chapter on 'Californian coal', despite the importance of coal for the power needs of the booming Californian gold mining industry. The influence of gold mining on the development of the regional power supply system in California is not considered. Rather than proposing gold as a theme for analysing electrification, these examples reveal close connections in the histories of electrification and gold mining. Another example is Hughes' portrait of Drexel, Morgan & Co. He presents this firm as the principal financier and marketer of Edison's electric light and power project, and he eclipses the private banking business enterprises of J.P. Morgan and his subsequent companies (Drexel, Morgan & Co. was succeeded by J.P. Morgan & Company in

---

mining companies, thereby courting speculative financial losses or gains on a grand scale' (Van Helten, 1982: 529).

158 These companies include, for example, the Edison Ore Milling Company Ltd. (1879), the Edison Iron Concentrating Company (1889), New Jersey and Pennsylvania Concentrating Works (1888), and Dunderland Iron Ore Company Ltd. (1902) (Rutgers Edison Papers).

1895), whose business involvement in electricity and gold is notable by any measure.

Viewed from the global perspective of the geographies of management and imagination, the private banking business enterprises of J.P. Morgan (and related companies) exemplify the high degree of correspondence between the business of electrification and the business of gold mining. J.P. Morgan and his companies acted as a financing agency for Edison's electric light and central power station project and some 70 years later (in 1948) issued a loan to the South African government to purchase and 'nationalize' the VFTPC. In between, he acted as the holder of rights to these Edison patents, stage-manager of the Edison technology in Europe (Paris Electrical Exhibition 1881, Holborn Viaduct power station) and in the US (Pearl Street Power Station), creditor for a private loan to the British imperial government to finance the Anglo-Boer War of 1899–1902 in South Africa and financier of Ernest Oppenheimer's Anglo-American Corporation in 1917 (initially founded as a gold mining company). Against this powerful record, the pioneering lighting of Drexel, Morgan & Co.'s Wall Street offices with power from Edison's prototype central power station on Pearl Street acquires new symbolic power.

Although it might be tempting to regard gold only as a powerful theme for the specific case of Johannesburg's regional power system, we may also consider it to be an example of the close association between electrification and the mining of natural resources more generally. From this perspective, the case of Johannesburg might show similarities with the early histories of electrification in other resource-based economies with a colonial legacy. The early international scope of American, German and British electrical manufacturing and consulting power companies (and the investment houses behind them) in places such as Argentina, Mexico and India illustrate the close fraternity between electrification projects and the imperial exploitation of natural resources. Furthermore, the specificity of the Johannesburg case found its way unfiltered into one of the grand textbooks of the day, written in 1913 by the leading engineer Georg Klingenberg from the A.E.G., and translated into English in 1916. This textbook was a significant opus in the establishment of the paradigm of large electric power systems – a paradigm that was still widely contested at the time.

Thus, as a theme, gold mining may serve to shed light on the entanglement of the history of electrification in Johannesburg with histories of electrification in other geographical places.

Empire

A white population of less than 1,200,000—about twice the population of Birmingham—has been entrusted with the duty of administering, under the Sovereign, the government of a country four times as large as the United Kingdom, with a trade of £80,000,000, a budget of £15,000,000, a debt of more than £110,000,000, a system of Government railways extending to more than 7000 miles; and the Union Government is to rule and look after the interests of more than 4,000,000 black and coloured people who (except in the Cape Colony) are not represented in the Assembly.
Sir Walter Hely-Hutchinson, 1911.

Hughes identifies a 50-year time span as the foundational years in the history of electrification, split into the years preceding and succeeding the First World War. Over these 50 years, the political world map underwent momentous changes. The years preceding the First World War correspond to a historical period often referred to as the New Imperialism. Hughes' history of electrification in Western society considers three cities in Britain, Germany and America over this period, without mentioning these countries' positions in this global political economy. His choice of countries, three major imperial powers, is by no means coincidental. Studying the history of electrification outside of these sites shows the intricate relationship between electrification and imperial history.[159] Moreover, this alliance did not set in

---

159 This text follows Miles' (2013) distinction between empire, imperialism, and colonialism: 'Notions of empire, colonialism, and imperialism have been afforded a range of meanings within different analytical frameworks and disciplines. [...] I use the term "colonialism" as a reference to explicit policies of formal territory acquisition and establishment of colonies. "Imperialism" has more informal implications, referring to policies that were not limited to formal colonialism, but pursued commercial and political expansionism, and involved the economic exploitation of target territories in circumstances

after the electric light and power station technology were invented and developed; rather, from the very beginning, this relationship was part of the project of developing electric light and power stations.

Empire does not only enter the stage after Edison's invention and development of the electric light and central power stations. Imperial objectives were part of Edison's early idea and project development: Edison deliberately chose Drexel, Morgan & Co. as a strategic point of entry to profitably position his (not yet invented) technology in the global political economy of the day. Revealingly, the first agreement between Edison and Drexel, Morgan & Co. regarding their rights to Edison's electrical inventions makes mention of the 'British dominions'. The sense of entitlement to privileged legal and economic rights in far-away territories is a key feature of the imperial period. Rather than simple 'technology transfer' across national boundaries of 'Western society', Edison's project plainly sought to create a world empire. This sense of entitlement rests upon assumptions of intellectual supremacy over other populations.[160]

The small lighting systems of the 1880s in Chicago, London, Berlin and Johannesburg evolved into regional power systems amid the national and industrial forces of a global political economy intent on empire building. The change in configuration of electric power systems during the half-century between 1880 and 1930 in these cities directly and indirectly forms part of these global dynamics. The three countries that Hughes used in his case study were the three most powerful nations at the time, and they fought for political and industrial territories across the globe during these years.

The Cape Colony, where the first Edison incandescent lights were installed in 1882, was a British colony, and the House of Assembly at Cape Town,

---

beyond actual annexation. The term "empire" is used to refer to the continuation of these imperialist practices, as well as to refer to the colonial empires created by Britain, France, Germany, the Netherlands, Spain, and Portugal, from the seventeenth to early twentieth centuries' (Miles, 2013: 4).

160 'Besides that, Americans and Englishmen, as Mr Fabbri said the other day, are a different kind of men to deal with from Continental people. I believe from my own experience that Americans are the honestest and most straightforward, as well as the best hearted people in the world. I think the English are next' (Edison Papers, Lowrey to Edison, 10. 12. 1878, D7821ZBR; TAEM 18: 226, pp. 3).

where the lights were installed, was an institution of British colonial rule. The first experimental use of electric lights in South Africa is recorded to have taken place in August 1881 in the Kimberly diamond mines – concurrent with the opening of the Paris International Electrical Exhibition, to which Hughes affords great significance for the technology transfer of Edison technology. These mines were established with mining rights to land that European settlers took by conquest. The objective of the mines was to serve the interests of imperial agents rather than those of the indigenous population. Indeed, the mines formed part of an intricate set of tools to gain control and power over the indigenous population.

Imperial interests were behind the first two power stations built in Johannesburg to supply gold mines; one was financed by the Deutsche Bank and the other by Rhodes' Consolidated Gold Fields Company. Power stations require certain rights and privileges for construction and operation. Obtaining such rights and privileges in settings that are controlled by European imperial powers is different from what is required to obtain them in places like Chicago, Berlin and London. In Johannesburg, the institutions and personalities that acquired these rights for the first two power stations shared a general sense of entitlement to appropriate and exercise rights and privileges in overseas territories. Sometimes this entitlement was conferred through public routes of passage, such as the purchase of land or coal. Even then, the public nature of the acquisition depended upon the subjugation of the rights and privileges of others for the transaction to proceed. At other times, the procurement of rights served to obscure grand ideas and goals, such as the imperial incentive of the Deutsche Bank to invest in RCEW of Goerz and Co. to establish a privileged point of entry for the German industrial manufacturer Siemens and Halske. In other cases, the rights and privileges were simply obtained through conquest – the land of the Transvaal republic had been conquered by the Boers and had only achieved the status of a self-governing republic less than two decades before its government issued the concessions to build and operate central power stations. Or, the rights and privileges were obtained by trade or wilful deceit by concession hunters, such as the German businessman Eduard Lippert.

The power station that Siemens and Halske built in 1897, the Rand Central Electric Works, was modelled after power stations that the company had built in Berlin. In Johannesburg, however, in addition to the power station, residential buildings, guest houses and accommodation had to be built for married and unmarried men, as well as manor houses with horse stables for the directors. This infrastructure was seen as urgently needed to recruit enough skilled professionals from Europe. The first power stations encouraged the importation of European settler communities to Johannesburg.

The founding of the VFTPC, which determined the subsequent course of electrification in Johannesburg and South Africa, also may be viewed as a result of the rival interests of the three imperial powers; Germany emerged as the winner, Britain was the loser, and America was left out because of Rathenau's agreement with General Electric in 1903 to divide the world market. After the First World War, this situation was inverted, and Britain and America took over the market previously dominated by German electrical technology.

Merz's recommendations on the electrification of South Africa in 1920 considered the national situation relative to the 'trend of opinion in Europe' (Merz & McLellan, 1920: 15). He viewed South Africa as a new country, a tabula rasa to be developed by white settlers. These 'imperial eyes' (Pratt, 1992) did not register the indigenous population other than as labourers but envisioned a future for the 'white' population only. Accordingly, his considerations for railway and electricity infrastructure only concerned this segment of the population and disregarded the needs of the rest of the population. Merz's recommendations led to the Electricity Act of 1922 and the establishment of Escom, which has survived in modified form to the present day as an effective monopoly over electricity generation, transmission and distribution in South Africa.

In sum, prior to the First World War, the small lighting systems of Johannesburg evolved into regional power systems on a wave of imperial conquest of Southern African territories. After the First World War, the monopoly of the regional power system in Johannesburg was endorsed by new regulatory authorities and legislation drafted by a British consulting company at the request of General Smuts. The recommendations of this consulting company were modelled on precedents in Great Britain, France, Germany, Sweden and the United States, and they promoted the establishment of monopolies and large regional power systems. Incidentally, this consulting company was also awarded the contract to supply and build the next large power station in Witbank.

Empire, imperialism and colonialism are essential themes for describing the changes in the electric power systems in Johannesburg from 1880 to 1930. They set the terms of the debate for electrification in Johannesburg and restricted the range of possible paths. They set the broad course for the electrification process in Johannesburg across all phases: Edison's global empire of patent rights including the territory of 'Southern Africa' (Edison Indian and Colonial Electric Company); the first lights and electric generators at the (foreign-owned) gold mines; the first two central power stations, owned by German and British companies; the first regional power scheme of the VFTPC, designed and built by the German A.E.G. and owned by German and British investment enterprises such as the Deutsche Bank and the British South Africa Company; and the British-designed national legislation and regulatory bodies established after the First World War to electrify the 'Unified' Republic of South Africa, which endorsed the monopoly of the foreign-owned enterprises of the VFTPC and ensured the growing market influence of British manufacturing and consulting businesses.

National competition between Britain, Germany and America

Three powerful nations competed for control of electrification in Johannesburg during the historical period under consideration: Great Britain, Germany and America. Electrification was a vehicle for enacting scenes of contest between these major competing powers on the global stage. Up until the First World War, Germany was the strongest competitor among these nations, but after the First World War, Great Britain took over the leading role, and America also entered the market. The competition among these three countries shaped the particular history of electrification in Johannesburg across all of its phases up until 1930.

The German company Siemens and Halske built and equipped the first power station in Johannesburg. It had already supplied telegraph equipment to the Cape Colony and stationed an agent in Cape Town before it established the Siemens and Halske South African Agency Johannesburg in 1895, and it was contracted to build the first power station in Johannesburg, the RCEW. The Transvaal was especially important to Germany at this time, and in 1896, it was described as 'the pivot of German imperialistic pretensions' (Penner, 1940: 58). German industrialists allied with the Boer government in the Transvaal. In a paper on Germany and the Transvaal before 1896, Penner notes that 'it was the hope of the German imperialists to create a second German Empire in South Africa, a second India under German control' (Penner, 1940: 31). By the early 20th century, Germany owned about one-sixth of the mining capital in the Transvaal (Weinberger, 1971). The close association between German banks and German government and industry reinforced the country's imperial objectives.

One route that was successfully pursued in this mission was to establish monopolies through the acquisition of concessions. For example, the German 'concession hunter' Eduard Lippert managed to secure concessions for the manufacture of dynamite, gunpowder and cement, the whisky monopoly was held by a German firm, and the concession for the National Bank and the South African railways were held by German–Dutch firms (Penner, 1940: 47). The first Transvaal government concession to establish a central power station was afforded to a German company for a mining company owned by the Deutsche Bank. In the mid-1890s, German electrical industries established export organizations in co-operation with German Banks (Deutsche and Dresdner Banks, Siemens and the A.E.G.). The concession to generate and supply power from the Victoria Falls to the Transvaal gold fields was purchased by the African Concessions Syndicate from the VFTPC, a company controlled by a syndicate of German banks and electrical industry, and the British South Africa Company.

A quote from the illustrious Emil Rathenau illustrates the extent to which electrification had to do with broader issues of competition between imperial nations. Rathenau wrote to the German Foreign Office to inform them about the impending conclusion of the deal between the A.E.G. and the British South African Company to establish the VFTPC.

This deal would represent an unexpected success, previously hardly achieved by German industry, and the large banks have made an effort [...] to make this business possible for interests more patriotic than material, because it stands without precedent in the electrical industry that an English Company would commission to a German company orders of such magnitude and thereby acknowledge its superiority over all other nations. The impartial print media has also duly honoured the success of German industrial competence and has made an effort to support the commenced work through factual statements. The undersigned respectfully had the honour personally to present the project to your majesty the emperor and to point out that the business relationships between the two nations are improving (Weinberger, 1971: 61[161]; my translation).[162]

German economic success in the Transvaal was a challenge to British hegemony (Penner, 1940): 'Germany was the natural ally of the Boer' (Penner, 1940: 53). London, the financial centre of the imperial world, figured as major hub for the registration of German, American and British companies generally and for electric utilities in particular. In the late 19th and early 20th centuries, Great Britain 'was the world's greatest capital exporter' (Wilkins, 1998: 5). Various kinds of joint-stock companies were established to serve as 'international capital flow conduits', and often these were foreign controlled (Wilkins, 1998: 9). Accordingly, American and British electrical industries were also present in South Africa from the very beginning. However, they were only able to overhaul the monopoly position of German electrical companies after the First World War. For example, the Anglo-American Brush Electric Light and Power Company (registered in London) established a subsidiary, the South African Brush Electric Light and Power Company, in 1882, but this company was discontinued after only few years.

The unchallenged position of London as the centre for most companies involved in the early electrification of the Witwatersrand is easily illustrated.

---

161 Weinberger (1971) is quoting DZA Merseburg, Rep. 120, CXIII 14, Nr. 46, Bd. 6, Bl. 107 f.

162 'Es verwundert daher nicht, mit welchem – durch ein "patriotisches" Mäntelchen kaum verhülltem – Triumph Emil Rathenau am 10. September 1906 an das Auswärtige Amt über den bevorstehenden Abschluss mit der Chartered Co. schrieb: „Dieser Abschluss würde einen ungeahnten, von der deutschen Industrie bisher wohl kaum erzielten Erfolg bedeuten, und weit mehr aus patriotischer Gesinnung als aus materiellen Interessen haben die uns befreundeten Grossbanken [...] um das Zustandekommen des Geschäfts sich bemüht, denn es steht in der elektischen Industrie ohne Beispiel da, dass eine englische Gesellschaft Aufträge in solcher Höhe freihändig an eine deutsche Firma überträgt und hiermit ihre Überlegenheit über alle anderen Nationen anerkennt. Auch die objective Presse hat diesen Erfolg deutschen industriellen Könnens gebührend gewürdigt und sich bemüht, durch sachliche Darstellungen das begonnene Werk zu fördern. Der ergebenst Unterzeichnete aber hatte die Ehre, Sr. Mjestät dem Kaiser persönlich über das grossartige Projekt Vortrag zu halten und darauf hinzuweisen, wie die geschäftlichen Beziehungen der beiden Nationen zueinander sich zu bessern anfangen' (Weinberger, 1971: 64).

For example, the Edison Indian and Colonial Electric Company had already used London as the base for its operations in 'the Colonies'.[163] The first two power stations, the RCEW and the GEPC, were both registered in London, the first manager for the VFTPC was recruited from the company Merz and McLellan in London and the headquarters of the Rhodesia-registered VFTPC was in London. Although London managed to retain its position as the financial centre of the world prior to the First World War, Germany and America dominated the electrotechnical market. Emil Rathenau's trip to America in 1903 to negotiate the division of the world market between the American and German electrotechnical industries was not presumptuous; it realistically reflected the clashing frontiers of the two global competitive players of the time. After the First World War, Great Britain took over Germany's prime position in the electricity manufacturing and consulting industry in Johannesburg.

The VFTPC, which was 'born of two empires', may be viewed as a joint British–German imperial project (Christie, 1984: 56). Renowned German, British and American engineers were summoned by various parties to provide expert judgements about the viability of the original project to supply and transmit electricity from the Victoria Falls to the gold mines of the Transvaal, including George Forbes of Great Britain, Georg Klingenberg of the German A.E.G. and Ralph Mershon of America. In the negotiations that preceded the establishment of the VFTPC and its subsidiary the Rand Mines Power Supply Company, local government, municipalities, business and civil society played no significant role. The future of electric power supply in

---

163   Registered in London in 1882, it was to act as parent company for Lighting and Power Companies in 'the Colonies' of Australsasia, South Africa, India, and Ceylon (Edison Papers, 1882: [D8239ZAO; TAEM 62: 827]). To 'enlist local capital and influence', this company would be to 'grant licenses to Corporations and local Companies' to '[light] up towns, public buildings, manufactories, barracks, and residences, and [supply] power by means of electricity' (Edison Papers, 1882:(D-82–40x); TAEM 63: 32, pp. 2). Edison appointed an agent in London as attorney to manage his letters patent of this company. The company intended to purchase and send out 'Edison's lamps, dynamos, and other materials which are now in London, ready for shipment to some of the principal centres of population and industry within the Company's field of operations' (Edison Papers, 1882: (D-82–40x); TAEM 63: 32, pp. 2).

Johannesburg was determined by the interests of German and British financial and industrial companies and their owners. The engineers for the
VFTPC also were recruited in England.[164] After the First World War, the
contracts for new power stations went to the British electrical and manufacturing industry.[165] Escom, in turn, was established in 1922 on the basis of
recommendations put forward by the eminent British consulting company
Merz and McLellan. These recommendations were modelled after British
legislation and the problems it sought to address. The subsequent negotiations that settled the relative powers and range of influence of the VFTPC
and Escom for the next 20 years were conducted under the personal media-

---

[164] 'In the attempt to ensure an "English" character for the VFTPC, not only Bernard
Price, but a large team of engineers had been recruited from the Tyneside power stations'
(Christie, 1984: 45).

[165] Bernard Price writes about 'Experience with German Firms and Plants': 'It may not be
out of place if I now refer to the Power Company's experience in dealing with certain
German concerns, because I do not think a better example could be cited of the methods
which our enemies have diligently pursued in their attempt to attain supremacy in the
industrial world. The Victoria Falls and Transvaal Power Company was promoted by British interests, and every effort was made to raise the necessary capital in Britain. Unfortunately, these efforts failed, and in the end certain German industrial banks took up debentures on condition that the main contracts were placed in the hands of German
manufacturing concerns with which they were allied. As a result, the A.E.G. obtained
important contracts on favourable terms. It is unnecessary for me to enlarge upon the
advantageous position in which the A.E.G. were thus placed but I would emphasise the fact
that the whole arrangement was the direct result of the German system of industrial banks,
under which financial assistance rendered for an industrial undertaking such as a power
company becomes the means of assisting German manufacturing firms. Needless to say,
this initial arrangement was not continued, and, as the power scheme grew and proved its
worth, capital was raised in London at the rate of no less than £1,000,000 sterling per
annum and quite independently from German banks. The A.E.G. then became faced with
keen competition, but this did not deter them in their effort to secure contracts for the
additional plant required. On the contrary, they at once reverted to the policy of dumping
their goods at low prices. [...] It must be remembered that, although feelings of sentiment
to-day run high, no purchaser in 1912 would have been prepared to sacrifice large sums of
shareholders' money in order to avert the dumping of German goods. In the end the
Germans secured most of the contracts at prices largely below those offered by their British
competitors' (Price, 1916: 63).

tion of Merz, and they resulted in contracts to British electrical and manufacturing industries.

In this way, national competition between Great Britain, Germany and America set the terms of the debate for the electrification of Johannesburg and restricted the range of possible paths to be taken. Competition between these countries was driven by notions of entitlement to establish large technological enterprises in overseas territories, which in turn required appropriating rights to foreign land and natural resources. Racial superiority over populations in far-away territories lay behind these claims.

Companies

At first view, the early history of electrification in Johannesburg appears to have involved a small number of companies only, such as the first two central power stations (the Rand Central Electric Works (RCEW) and the General Electric Power Company (GEPC)) and the company that established the first regional power scheme, the Victoria Falls and Transvaal Power Company (VFTPC). However, it developed under the financial, administrative and legal patronage of a large number of companies, including the Edison Indian and Colonial Electric Company, the Deutsche Bank, Consolidated Gold Fields of South Africa, Siemens and Halske, the A.E.G., the British South Africa Company, the Corner House group of companies of Beit and Eckstein, the Rand Mines Ltd., N.M. Rothschild & Sons, the Exploration Company Ltd., the Central Mining and Investment Corporation Ltd., and others. The various labels given to these enterprises include the private or public limited liability company, the joint-stock company, the free-standing company, the holding company and its subsidiaries, the investment group, the chartered company, the merchant firm, and others. Needless to say, these companies spanned a broad spectrum of objectives, strategies and dealings that are typically not accessible through a given type of company. However, from the viewpoint of Johannesburg, most of these companies had three attributes in common: they were foreign owned, foreign managed and foreign financed. Another striking feature is their perpetual transformation. Over time, their continually changing status, structure, operations and focus produced a 'forest of company names and [...] jungle of business groups' (Hausman et al., 2008: 41).

Despite the complexity of companies involved, a few features emerge in the early history of electrification in Johannesburg. Several companies were established by German initiatives and formed part of Germany's strategy of economic imperialism in the Transvaal. German-owned or co-owned companies such as Goerz, and the RCEW and the VFTPC were financed by German banks or by banking syndicates that were largely German. Often these companies were registered in England, and the German ownership was not disclosed to the public. Another group of companies were British owned and financed, such as the GEPC, Consolidated Gold Fields, the BSAC and the African Concessions Syndicate. These companies were more openly connected to British imperial economic and political interests in Southern Africa. However, a fine line separates 'British' from 'South African' attributes, because the status of the Transvaal kept changing over the years 1880 to 1930: this area was once considered to be an independent Republic (but still under British suzerainty), then it was a British colony, and after the union of the four colonies (Cape Colony, Natal Colony, Transvaal Colony and Orange River Colony) in 1910, it became a non-sovereign dominion of the British Empire.

In countries with local electrical manufacturing industries and local capital, the companies involved in the business of power stations are easily traced and profiled. In Johannesburg, the companies that appeared at centre stage in electric power projects were entangled in a complex and often unintelligible labyrinth of pursuits and ideas because of their dependence on foreign capital, foreign expertise, foreign legal provisions, foreign management and foreign national policies. These intricate dependencies cannot be simply disentangled by tracing the technology, the economic, social setting or policies of electrification. This complexity is a challenge for historical analysis. However, it granted the companies an ideal environment to undertake strategic business manoeuvres.

During the first phase of Hughes' history of electrification, an American-owned company registered and managed in London claimed rights and privileges in a vaguely indistinct area referred to as South Africa. The Edison Indian and Colonial Electric Company was part of the family of Edison companies that were established between 1880 and 1883 to exchange patent rights for equity interests domestically and abroad. At the time, companies were often registered under British law.[166] Although the Edison Indian and Colonial Electric Company, along with most other companies connected to Edison's electric light and central power station project, became independent within a decade, these business enterprises drove the early international diffusion of Edison's technology, and therefore may be classified as a multinational enterprise group (Hausman et al., 2008: 77).

During the second phase of electrification, large electrical companies such as the German A.E.G. and Siemens, the American General Electric and Westinghouse, and the Swiss Brown Bovery, were carrying out international projects, and for this purpose, they established manufacturing and sales enterprises (Hausman et al., 2008: 92). For example, in 1895, by the end of this phase, Siemens and Halske founded a South African Agency with headquarters in Johannesburg and, together with the Deutsche Bank, established a joint agency to promote their commodity exports in South Africa (the Technical and Commercial Corporation Ltd.). The German electrical companies assumed diverse roles in these years, and operated as finance institutions, manufacturers and energy suppliers.[167]

Segreto notes that the 'expanding market for electricity also brought new kinds of complications to the companies involved in this field, particularly the

---

166   Wilkins proposes five categories of joint-stock companies that were set up under British joint-stock company law and operated as 'international capital flow conduits' as part of multinational enterprises or under the control of foreign individuals (Wilkins, 1998: 9). Overseas investment occurred through domestic companies (developed business operations abroad after registration); domestic and international companies (with domestic and international business at the time of registration); trading and shipping companies, investment trusts and other financial intermediaries, and free-standing companies (organized to undertake business in foreign countries or regions) (Wilkins, 1998: 8).

167   The authors quote the German economist Jacob Riesner: "'[T]here was scarcely a form of management or financing which was not utilised in the nineties by the [German] electrical industry. There were syndicates, subsidiary companies (*Tochtergesellschaften*), and trust companies [...] operating companies proper, and manufacturing companies, as well as financing institutions, increases and reductions of capital, silent participations, commandites, issues and sales in the open market, fusions, pooling of profits (*Gewinngemeinschaften*), buying of shares, separations and combinations, independent and syndicated enterprises at home and abroad. In short, a medley of undertakings [...]'" (Hausman et al., 2008: 95).

electrical equipment manufacturers' (Segreto, 1994: 163). The management of electric companies required a costly professional bureaucracy, and these companies employed various strategies to maintain and expand their markets (Segreto, 1994). The holding company was preferred by all of the major equipment manufacturers over limited liability or joint-stock companies.[168] The holding company, the equipment manufacturer, and the electric operating company were associated by contract and were represented in each other's respective boards of directors.[169] Hausman and colleagues consider holding companies to be 'a pyramided corporate structure that came to be an umbrella designation for highly diverse underlying behaviour' and as a 'very important set of initiators in spreading electricity around the world' (Hausman et al., 2008: 52). The case of Johannesburg reveals the complicity of these companies with German imperial ambitions.

During the third phase of electrification in Johannesburg, the first central power stations, the Rand Central Electric Works (RCEW) and the

---

168 According to Segreto, the relationship between the electrical equipment producers and their financial holding companies came to be known as *Unternehmergeschäft*: 'The financial trusts' purpose were to establish and finance, alone or, more often, in collaboration with other partners, companies for the production and distribution of electricity; they planned the projects and supervised the construction of the plants, just as a modern engineering consulting firm would, obtaining from that activity a commission of approximately 7 % of the total costs of construction. The machinery and the necessary equipment to run the power plants were provided by the electrical equipment manufacturer connected to the holding company or by the manufacturer's affiliated companies. Once the electric companies were able to distribute dividends regularly, the holding companies completed their job by entering the shares on the market' (Segreto, 1994: 164).

169 'There were numerous connections among the three actors – that is, the equipment manufacturer, the holding company, and the electric operating company. Representatives of the equipment manufacturing firms as well as those of the participating banks were among the directors of the holding companies. Representatives of the holding companies sat on the boards of directors of the electric power companies that the holding companies and the banks financed. The relations between each electrical equipment enterprise and its holding company were defined by contract and bound the choices of the latter to the commercial needs of the former. However, the holding companies had quite a wide margin of autonomy from the head office, especially in their distinctly financial operations, and sometimes, as in Indelec's case, moments of tension and dispute occurred' (Segreto, 1994: 164).

General Electric Power Company (GEPC) were established by two groups of powerful companies. A brief sketch of these companies illustrates the variety of objectives, structures and operations that were subsumed under the category of the 'company'. At first glance, the RCEW and the GEPC appear to be simply connected to the mining companies of their founders; Goerz and Co. (founded by the German Adolf Goerz), and the Consolidated Gold Fields Group (formed by the Britons Cecil Rhodes and Charles Dudd). However, the RCEW and the GEPC branched out into a complex group of diverse but interrelated business ventures. These business ventures display an opaque diversity of organizational forms and objectives. Notably, the legal provisions governing the international business of these companies at the end of the 19th century were not subjected to international company law. Rather, the laws of selected European nations determined the legal framework for territories under colonial rule. One of the model legal provisions for these laws was the British Joint Stock Companies Act of 1856 which introduced the principle of limited liability.

The RCEW illustrates an industrial business venture in which banks and industry collaborated closely to further the interests of the German Empire in the Transvaal. It presented itself in public as founded and owned by Adolf Goerz and his company Goerz and Co. However, the RCEW was initiated and financed by the Deutsche Bank and equipped and built by the German company Siemens and Halske. The private company Goerz and Co. (established in 1892), too, was controlled by a syndicate of German banks: the Deutsche Bank had sent Goerz out to South Africa in 1888 to explore business opportunities and had formed a syndicate to provide the capital of 100,000 pounds to found Goerz GmbH (the predecessor to the private company Goerz and Co.). Incidentally, Adolf Goerz's sister was married to Georg von Siemens, the founding director of the Deutsche Bank and nephew of Werner von Siemens (co-founder of Siemens and Halske).

The GEPC, on the other hand, was built by Consolidated Gold Fields of South Africa, which formed part of a family of companies under the influence of Cecil Rhodes and Alfred Beit. Rhodes had also been involved in the establishment of the De Beers Consolidated Mines in Kimberly and the British South Africa Company, all of which had been supported by N.M. Rothschild & Sons, a British investment banking house. The Consolidated Gold Fields of South Africa also was associated with the 'Corner House' Group of companies established by the German financiers Beit and Eckstein. These connections merged financial and business interests with political power and imperial visions, and paved the way for the large electric power scheme that was built in Johannesburg before the Second World War.

The holding company *Rand Mines Ltd.*, established in February 1893 by Beit and Eckstein, illustrates the measure of control exerted by these business associations. Rand Mines Ltd. was a group mining finance house that provided 'a centralised set of managerial, secretarial and administrative services' to facilitate the provision of (predominantly European) investment capital (Turrel & Van Helten, 1986: 188).[170] The group of investors involved in its establishment included the Consolidated Gold Fields of South Africa and the *London Exploration Company Ltd.*, which had been established in 1886 by N.M. Rothschild & Sons and other merchant bankers. The original purpose of the Exploration Company was 'simply [to] assess mining propositions and recommend investments to its members'. This brief was changed to that of company promotion in 1889 when the company was restructured as a joint-stock venture (Turrel & Van Helten, 1986: 184). The Exploration Company 'combined engineering and financial expertise' (Turrel & Van Helten, 1986: 182) and recruited the American mining engineers Hamilton Smith and Edmund de Crano to run the company. The Rand Mines Ltd. was a 'giant holding company' and 'undoubtedly the most profitable concern on the Reef' (Turrel & Van Helten, 1986: 188).

The establishment of Rand Mines Ltd. has been described as a 'turning point in the system of capital provision'[171] (Graham, 1996: 6). Incidentally, it

---

170   The first general manager of Rand Mines was an American engineer, Henry Cleveland Perkins, who had previously served as consultant to the Exploration Company (Turrel & Van Helten, 1986: 188). Two directors of the Exploration Company were also directors of the BSAC (Sir Horace Farquhar and Rochfort Maguire) (Turrel & Van Helten, 1986: 188).

171   This new style is described as the 'group system': 'In the first instance, it signalled the start of the "group system", whereby each mine was controlled by one of a few huge groups. Each mine was floated as a joint stock company with its own directors and its own manager, the Group maintaining control through share ownership and, more importantly, by dominating the board of directors. The individual companies were never wholly-owned subsidiaries, but rather the objective was to raise capital by flotation and the

involved two powerful friends, Cecil Rhodes and Alfred Beit: '[...] through thick and thin Rhodes retained valuable support from Alfred Beit and his partners, H. Eckstein and Julius Wernher, their agents on the Rand, and from their bankers, N.M. Rothschild & Sons. It is important to emphasize that the Johannesburg subsidiary of the Wernher, Beit house, namely H. Eckstein & Company, was in fact run by subordinate employees who were supportive of Rhodes's imperial schemes and deeply implicated in the Jameson Raid' (Newbury, 2009: 98). Consolidated Gold Fields was one of two sources of Rhodes' financial power that allowed him to pursue his programme of political expansion in Africa (Galbraith, 1975: 57) (the other being De Beers).[172]

By the end of this phase, the sixty-six gold mines of the gold mining industry in the Transvaal in 1905 covered a broad variety of corporate, management and ownership structures and were often registered under the legal authority of foreign sovereign powers. They operated as finance, exploration and trust companies under the auspices of distant legal and political authorities and continually changed their structure.[173] In addition, joint-stock

subsequent sale of vendors' interests, and in many cases to make substantial profit by the sale of shares, once stock prices had risen' (Graham, 1996: 7).

172   'The most important controller of the mines, Julius Wernher, the financial genius behind the Corner House, was of course an international entrepreneur who directly controlled local capital allocation and development tactics from his vantage point in the City. Both he and most of the other mining magnates sought funds from all of Western Europe's primary and secondary markets. Through stock market operations, the entrepreneurs siphoned large amounts of capital into their personal fortunes. [...] Thus, the strategies of these two mining houses [Corner House, Goldfields] served to amass significant amounts of wealth through gold mining finance which were then invested, neither in the gold mining industry nor in South Africa, but elsewhere in the world. [...] Though the mines' controllers may be variously labelled, it was their function as international developers and speculators which was fundamental to the evolution of the industry' (Kubicek, 1979: 196–7).

173   For example, during the early 1890s, as mines increasingly had to dig deeper to recover the ore and as extraction and processing techniques became more costly, there was a tendency to subsume several territorial claims for gold mines into holding companies to provide the necessary capital for larger scale investments. The change in company structure and status of the enterprises established by the German Adolf Goerz, who also founded the

companies came to function as intermediary financial advisors to investors, such as the London Exploration Company Ltd.

At the beginning of the fourth phase of electrification, Wernher Beit & Company established another financial investment company in May 1905, 'the biggest trust of its kind the Rand or even London and Paris had seen' (Fraser, 1975: 166, quoting Kubicek, 1979),[174] the London-registered *Central Mining and Investment Corporation Limited,*[175] with a capital of six million pounds. According to Kubicek, more than one-third of the Witwatersrand's gold output between 1902 and 1913 (which amounted to 11 % of world output) 'was produced by companies controlled by the Central Mining and

---

RCEW in 1897, reveals the nature of the investment business involved during this phase. What started out as a syndicate Goerz GmbH, was established as Ad. Goerz and Co. Limited under German Law in 1889 and restructured as A. Goerz and Company, Limited in 1897, 'with a view principally to finance, exploit, develop, and work gold mines in South Africa and elsewhere' (Skinner, 1911: 141). By the end of the year 1903, 'the company held 512 claims, most of them being well-situated deep-level blocks in the western district of the Rand, and its land holdings consisted of 7,480 acres of unproclaimed deep-level ground in the Western and Eastern districts, including the western half of the farm Witpoort, upon which five bore-holes have cut the main reef' (Wills & Barrow, 1905: 275). In June 1904, the capital was increased to present amount [1,500,000 pounds], [...] the Deutsche Bank guaranteed the entire issues for 1 s. per share [...] (Skinner, 1911: 141).

174 'Wernher Beit's own resources were likely so large that it need not have called upon outside sources or close associates for the additional capital the deep-level mines required. Securities the firm had deposited with the Union Bank of London in 1895 are instructive. On face value these securities were worth £1,170,000. Almost half this amount, or £547,000, was in Consols and earning 2 3/4 percent. Bonds issued by the governments of Argentina, Chile, Cape Colony, the Transvaal, Egypt, Russia, and India had a value of £233,000. American securities, mostly railway bonds, totaled £230,000 and promised an average yield of 4.8 percent. [...] Other small commitments included Netherlands and South African Railways debentures (£50,000), Beira Railway bonds (£23,000) [in Mocambique] and Fraser and Chalmers securities (£5300). [...] it was increasingly becoming a London-based private investment house with diversified interests overseas' (Kubicek, 1979: 68–9).

175 'Central Mining was formed originally to take over the business of the African Ventures Syndicate which Wernher had formed in 1903 to buy old shares [...] and to attract mainly French capital to finance the mines during the reconstruction of the Transvaal, after the South African war' (Fraser, 1975: 166).

Investment Corporation, by its affiliate, Rand Mines Limited, or by the creator of these two investment trusts, the private financial house of Wernher, Beit and Company' (Kubicek, 1979: 53). The decision of this gold mining imperium to convert their mines to use electric power determined the course of history of electrification in Johannesburg and South Africa. It became the major consumer of electric power in Johannesburg and the principal customer of the VFTPC.

The VFTPC was registered in Rhodesia, but it presented itself as a British company, with head offices in London and one million pounds of ordinary shares belonging to the African Concessions Syndicate Ltd. (which was, in turn, owned by the BSAC). However, of the two million pounds of preference shares, 900,000 pounds were owned by the A.E.G. and the consortium of German banks. All debentures were owned by the A.E.G. and the consortium of German banks. Both the chairman and the managing director were British (Marquis of Winchester, A.E. Hadley), with ten British and two German (Emil Rathenau, Hans Schuster) members on the board of directors. The German A.E.G. engineer, Georg Klingenberg, was on the technical advisory board. According to Weinberger, this form of association complied with the corporate policy of the A.E.G. (Weinberger, 1975: 63). Of note are the 15 banks initially involved in the German finance consortium, which also include several Swiss banks: Bank für Handel und Industrie; Berliner Handelsgesellschaft; Disconto-Gesellschaft; Dresdner Bank; Gesellschaft für elektrische Unternehmungen; Nationalbank für Deutschland; Schaffhausenscher Bankverein; S. Bleichröder; Delbrück Leo & Co.; Hardy & Co., GmbH, sämtlich in Berlin; E. Heimann, Breslau; Gebrüder Sulzbach, Frankfurt a.M.; Bank für electrische Unternehmungen, Zürich; Schweizerische Kreditanstalt, Zürich und Basel; Aktiengesellschaft vorm. Speyer & Co., Basel (Weinberger, 1975). The A.E.G. had already established a bank syndicate under the leadership of the Deutsche Bank, the Deutsch-Überseeische Elektrizitätsgesellschaft, with a capital of 10 million Deutsche Mark, to establish electric installations and to purchase and finance enterprises in the field of applied electricity (Sothen, 1915: 65).[176]

---

176   'Um besonders in Südamerika festen Fuss zu fassen, gründete die A.E.G. im Jahre 1898 mit ihrem Bankkonsortium unter Führung der Deutschen Bank die Deutsch-

The other major shareholder of the VFTPC, the British South Africa Company, had been incorporated under British law by Royal Charter, 29 October 1889, instigated by Cecil Rhodes, with an original share capital of one million pounds. Its 'principal field of operations' was defined as 'in that region of South Africa lying to the north of Bechuanaland and to the west of Portuguese East Africa'. The BSAC's legal powers are elusive; the company was 'specially authorized and empowered' to issue shares, borrow money, make loans, establish banks and companies, to 'make and maintain roads, railways, telegraphy, harbours', to 'carry on mining and other industries and to make concessions of mining forestall or other rights', to 'improve, develop, clear, plant, irrigate and cultivate' land, to 'settle [...] territories', to 'acquire and hold personal property', and to 'carry on any lawful commerce, trade, pursuit, business, operations, or dealing' – in short, to 'do all lawful things incidental or conducive to the exercise or enjoyment of the rights, interests, authorities and powers of the Company in this Our Charter expressed or referred to, or any of them'. The Charter states:

> That the Petitioners desire to carry into effect divers concessions and agreements which have been made by certain of the chiefs and tribes inhabiting the said region, and such other concessions agreements grants and treaties as the Petitioners may hereafter obtain within the said region or elsewhere in Africa, with the view of promoting trade commerce civilization and good government (including the regulation of liquor traffic with the natives) in the territories which are or may be comprised or referred to in such concessions agreements grants and treaties as aforesaid.[177]

The line and power of jurisdiction for legal disputes relating to the BSAC are worth highlighting: 'In case at any time any difference arises between any chief or tribe inhabiting any of the territories aforesaid and the Company, that difference shall, if Our Secretary of State so require, be submitted by the

---

Überseeische Elektrizitätsgesellschaft mit einem Kapital von 10 Millionen Mark. Zweck der Gesellschaft ist Bau- und Betrieb elektrischer Anlagen, Erwerb und Finanzierung von Unternehmungen auf dem Gebiete der angewandten Elektrizität, insbesondere Beleuchtung, Transportwesen usw.' (Sothen, 1915).

177 Charter of the British South Africa Company, (London Gazette), 20 December 1889.

Company to him for his decision, and the Company shall act in accordance with such decision.'

The BSAC's function was to assume the risk of extending the infrastructure of modern capitalism (including railways) into south-central Africa for the benefit of the British, but without the British taxpayer having to pay the costs.[178] Unlike normal companies, the BSAC was permitted to establish a political administration with a paramilitary police force in areas where it might be granted rights by local rulers. It was also allowed to profit commercially through its own operations or by renting out land, receiving royalties on the mining of minerals, levying customs duties and collecting other fees.

One of the shareholders in the BSAC, in turn, was Wernher, Beit & Company. This company led the Corner House group, which by 1913 controlled 15 Rand mines 'with an issued share capital of more than £14 million' (Harvey, 1989: 115). At the same time, these mines were the main customers of the VFPTC's subsidiary, the Rand Mines Power Supply Company. All of these companies operated under a self-awarded sense of entitlement, appropriation and intellectual ascendancy that may be encapsulated in their designation as 'settler companies'.

The above image contrasts with Hughes' conclusion that the institutional form of a public and private utility 'presided over' and was influenced most directly by electric power systems (Hughes, 1983: 15). In Johannesburg, a diverse set of companies that prima facie appear unconnected to the business of electrification, in fact presided over its development. These companies offer a window into several underlying forces in the history of electrification in Johannesburg between 1886 and 1930. These forces connect

---

178   'The sub-Continent is particularly indebted to joint-stock enterprise. Even Rhodesia, a country larger in area than France, Germany, Austria and Italy combined, is administered under what is for all intents and purposes the Joint Stock Companies Acts. At a time when more than one European Power was anxious to establish itself in Africa, the British Imperial Parliament could not undertake the vast responsibilities involved in the acquisition of such an extensive territory as that which has for years borne the name of Rhodesia; and had it not been for the foresight and patriotic enterprise of Mr. Cecil Rhodes and his associates in the formation of the Chartered Company, Matabeleland and Mashonaland would probably have fallen to either one of these Powers, or would have become part of the South African Republic' (Wills & Barrett, 1905: 257).

directly to the imperial politics and industrial policies of America, Germany, and Britain. The companies involved in the establishment of the first power stations of Johannesburg and the subsequent grand scheme of the VFTPC were foreign companies. They pursued numerous purposes, including the investment of capital, the administration and management of electrification projects, the sale of manufacturing equipment, the generation, marketing and sale of electricity, the acquisition of land and rights, the establishment of living quarters and settlements, territorial hegemony, racial and social control, national imperial interests, personal careers, and the implementation of ideologies of modernization and civilization. In legal terms, these companies operated in the no man's land of emerging international law. Most of them were registered in Great Britain and therefore fell under that country's business laws. However, the degree of power formally bestowed and actually executed by these companies, their privileges, spheres of operation and influence are not readily appreciated.

An illustrative example is the A.E.G., the company that equipped, designed and built the gigantic regional power scheme in Johannesburg. Emil Rathenau travelled to America in 1903 to negotiate the division of the world electricity market with General Electric on behalf of the A.E.G. However, Rathenau entered into these negotiations as the director of a global business empire, not as the director of a German manufacturing company:

> This time he did not make his appearance in the New World as one who seeks to assimilate a small part of their accumulated richness of intellect to take it home to establish a humble livelihood. He made his appearance as intellectual sovereign, as industrial king, who intended to confront the leading men over there on equal footing to negotiate the distribution of the electrical world (Pinner, 1918: 269).[179]

---

179 'Diesmal erschien er aber in der Neuen Welt nicht als einer, der einen kleinen Teil des drüben angehäuften Geistesreichtums in sich aufnehmen und zur Errichtung einer bescheidenen Existenz im Heimatlande mit sich forttragen wollte, sondern als ein Geistesherrscher, ein Industriekönig, der den führenden Männern drüben als Gleichberechtigter entgegenzutreten und mit ihnen über die Verteilung der elektrischen Welt zu verhandeln beabsichtigte' (Pinner, 1918: 269).

Most of the companies that were involved in the early history of elec-trification in Johannesburg were legally accountable to overseas nations and pursued their imperial interests. These conditions generated complex legal situations in which "'legal variations" and parallel legal orders co-existed within the same territory' (Miles, 2013: 26).[180] Companies both contributed to and made extensive use of the ambiguous legal circumstances. They set the terms of the debate for the electrification of Johannesburg by obscuring a variety of underlying interests.

Concessions

The legal garment clothing a concession may vary.
Peter Fischer, *Encyclopedia of Public International Law*, 1985.

It is not in fact possible to exclude a priori any subject matter from the possible sphere of a concession. By its nature the State is capable of assert-ing an exclusive competence in any sphere of activity not denied to it by a rule of international law, as a slave trade concession would be today.
Peter Fischer, *Encyclopedia of Public International Law*, 1985.

The supply of electric power is both a legislative and a technical question.
Charles Merz & William McLellan, 1920.

Electric lighting and power schemes require the purchase of certain rights and privileges, such as for wayleaves, water, coal, land, transmission poles, and for the general operation, generation and sale of electric power. Acquisi-tion of such concessions is part of the traditional trade of electric power undertakings. In Hughes' history of electrification in Western society, such rights and privileges are referred to as patents, franchises, licences and con-cessions. Typically in his narrative, they were granted by local or national government agencies and fell under their respective legislative and regulative sovereignty. Overall, Hughes views concessions, licences and franchises as

---

180   Quoting Benton, Lauren. 2010. *Search for Sovereignty. Law and Geography in Euro-pean Empires, 1400–1900*. Cambridge, New York: Cambridge University Press, pp. 2–8.

forming part of the *local* legal and political process required to establish power systems. For the case of Johannesburg, the rights and privileges conferred through concessions reach far beyond the local, and they also transcend legal and political matters.

Hughes first mentions concessions in connection with Edison's plans to sell 'concessions or licenses' to prospective companies in the United Kingdom through English Edison (Hughes, 1983: 55). As early as 1882, Edison claimed patent rights in South Africa with the establishment of the Edison Indian and Colonial Electric Company. Except for a few lighting projects, however, these rights were not used. In the Transvaal Republic, rights of local authorities to establish electricity undertakings were only established in the 1890s, following the model of legislation in Great Britain, where the Board of Trade regulations regulated municipal electricity supplies (Marquard, 2006: 144). However, diamond and gold mines continued to install electric lights and (later) small electric power plants without obtaining government permission. The Johannesburg Gas Works did not need to obtain a concession for their plant. When it was commissioned, local government was still new. The first government of Johannesburg was founded in 1886, and it took the form of a 'Digger's Committee', which consisted of nine members under the chairmanship of the Mining Commissioner. Within a year, it was succeeded by the Sanitary Board, which acted as the local authority and was also headed by the Mining Commissioner.[181] The Sanitary Board, in turn, was replaced by a Town Council in December 1897, when the status of Johannesburg was changed to that of a town.[182] When the British occupied Johannesburg during the Anglo-Boer War in 1900, the town became an 'Imperial Government Municipality' and the Military Governor was assisted by an Acting Mayor who had to administer municipal affairs until it was possible to establish a

---

181 The Sanitary Board originally included five elected members and two government nominees. Two years later, it was reconstituted 'to consist of twelve elected members under the chairmanship of a Government Commissioner and its area of jurisdiction was five square miles' (Public Relations Office, 1967: 2).

182 'The Town Council comprised twelve elected members and a 'Burgomaster' appointed by the Government. The Council was given powers to make regulations in regard to safety, public order, morality and health' (Public Relations Office, 1967: 2).

civil form of local government in 1901.[183] Thus, concessions in Johannesburg during this period were granted under the unstable circumstances that are typical of territories under the fierce contest of imperial powers.

Significantly, the first government concession to establish a central power-er station was awarded to a foreign company, the German Siemens and Halske, in July 1894. This concession was drafted in the form of an agreement signed by the state secretary and a representative of Siemens and Halske – the operating company (the Rand Central Electric Works (RCEW) owned by the gold mining company Goerz & Company) was only established two years later. The concession included 'the right to lay and have electric conductors for the transmission of power to the mines along the Witwatersrand' on certain conditions.[184] The concession to establish the second power station in Johannesburg was acquired in 1897 by the Simmer and Jack Proprietary Mines Ltd., a subsidiary of the Consolidated Gold Fields Group, and it was signed by Jan Smuts in his capacity as Registrar of Deeds. It was ceded to the General Electric Power Company (GEPC) in 1899 and conferred 'the right of laying and maintenance of electrical installations for motive power' on behalf of six mines owned by the Consolidated Gold Fields Group (Power Companies Commission, 1910: 52). At the time, the Transvaal was a self-governed, independent republic under British suzerainty, which meant that it actually was not entitled to enter into agreements with foreign countries without the consent of the government of the Great Britain. The Siemens and Halske concession was granted with the option of expropriation by the government after 15 years. Both concessions specified that 'amalgamation with other grants for power transmission can only occur with the approval of the Government' (Power Companies Commission, 1910: 48, 54).

---

183   In 1901, Lord Alfred Milner became Administrator of the Transvaal and invited twelve prominent citizens to serve as town councillors until such time as proper elections could be held' (Public Relations Office, 1967: 2).

184   'The Government grants to the firm Siemens and Halske in as far as such is not in conflict with rights already granted, the right to erect poles and to fix electrical conductors to the same on Government grounds, public roads, and proclaimed goldfields [...]' (Power Companies Commission, 1910: 47).

Despite these specifications, both concessions were ceded to the newly established VFTPC in 1907. As part of its strategy to achieve a monopoly to supply electricity to the gold mines of the Witwatersrand, the VFTPC first obtained the necessary concessions to purchase and extend existing undertakings. This included the purchase of the Lewis and Marks concession in 1905 for a pole line from their coalfields at Vereeniging to supply power to the Witwatersrand 25 miles to the north as well as 'an agreement for the right to establish a power station at Vereeniging' (Hadley, 1913: 3). The company from which the VFTPC purchased this concession belonged to cousins Isaac Lewis and Sammy Marks, who had previously considered supplying the Rand from their coalfields at Vereeniging, 25 miles south of Johannesburg, and they had obtained wayleaves for a pole line for this purpose (Hadley, 1913: 3). Sammy Marks was a friend of President Kruger.

The RCEW and the GEPC were built on land classified as farmland. The first power stations in Johannesburg required rights to use this land and water, and rights to build and maintain transmission lines to distribute electricity. The concessions for the RCEW[185] and the GEPC[186] were issued by the

---

185 Siemens and Halske obtained a concession on 19 July 1894 and ceded it to the RCEW in 1895 '[...] the right to lay and have electric conductors for the transmission of power to the mines along the Witwatersrand' (Chamber of Mines, 1911: 45). The government grant stated several conditions, including the restriction that 'amalgamation with other grants for power transmission can only occur with the approval of the Government' (Chamber of Mines, 1911: 46). This concession was amended on 5 February 1906 to cede the rights of Siemens and Halske to the Rand Central Electric Works Ltd.

186 The Simmer and Jack concession was obtained on 31 July 1897 and ceded to the GEPC on 16 March 1899. Simmer and Jack Proprietary Mines Ltd. were granted 'the right of laying and maintenance of electrical installations for motive power on behalf of the Simmer and Jack Proprietary Mines Limited, Simmer and Jack East Limited, Simmer and Jack West Limited, Knights Deep Limited, Jupiter Gold Mining Company Limited, Rand Victoria Mines Limited. This included 'the right to erect poles on Government lands, public roads, and proclaimed mining areas [...] for the transmission of motive power' (Chamber of Mines, 1911: 52). Again, it was stipulated that 'amalgamation with other grants for motive power can only take place with the approval of the Government' (Chamber of Mines, 1911: 54). The concession was amended on 14 November 1905 to grant the GEPC 'the right to erect poles and to affixt hereon electric mains or conductors

South African (Transvaal) government, which had gained independence from Great Britain in 1884. The rights of the Transvaal government to issue these concessions regarding land had been granted by Great Britain in the London Convention of 1884. The indigenous population of this territory was not considered in the institution of this European land ownership scheme. The land of the Transvaal had been contested for centuries. The early African land tenure practices of the Africans were displaced by various populations. By the end of the 19th century, all African populations had been conquered. In other words, the concessions were issued by a Boer government with political authority conferred by an imperial European power, and not by the local population.

However, the concessions granted to the RCEW and the GEPC were not comparable to the concessions or franchises granted by municipal or national governments for the construction and operation of the first central power stations in Berlin, Chicago or London at the time. In Great Britain, legislation to regulate the excavation of streets for electrical lines was passed by a political process that ended within two weeks of Edison's model central power station opening at Holborn Viaduct. The electric lighting bill was developed by the Board of Trade,[187] which also dealt with transportation, communications and industries, and therefore could draw from precedents concerning, for example, tramways or waterworks. This legislation was supported to protect the public against the tendency of public utilities to abuse monopolistic powers. In Berlin, Emil Rathenau's Deutsche Edison-Gesellschaft, having acquired patent rights from Edison, established the Städtische Electricitäts-Werke in 1884 and entered into a concession agreement with the municipality of Berlin to supply it with electric power.

The strategy pursued by the German company Siemens and Halske, however, was not unique among multinational electric power business

for the transmission of electric power over Government and private ground, public roads and ways, proclaimed goldfields' (Chamber of Mines, 1911: 55).

187   Incidentally, the Board of Trade during the time of this legislation was presided over by Joseph Chamberlain in Gladstone's cabinet, from 1880 to 1885, who later assumed the post as Colonial Secretary to British Colonies from 1895 and was accused of complicity in the Jameson Raid.

ventures. From the beginning, they included 'the scouting and negotiation of the franchise, the securing of the contract, engineering, company promotion, underwriting, construction, operations and management, and primary and secondary distribution of securities' (Hausman et al., 2008: 47). In colonial contexts, which are characterized by a multiplicity of unstable and ever changing governance and administration structures and rules, such business enterprises profited from the confusing, transitory and unprotected local conditions. Because of the types of conditions that prevail when a region is under the siege from imperial powers, the legal designation of a concession in a place like Johannesburg assumes a broader spectrum of meanings than it would in places like Berlin, Chicago and London.

For example, the concession acquired by the Consolidated Gold Fields Group (through its subsidiary the Simmer and Jack Mines) to establish the GEPC was part of a concession-hunting policy that Cecil Rhodes (and his companies) used to further his imperial vision of expanding the British Empire. In fact, when the concession for the right to supply electricity to five mines was issued in 1897, Rhodes had only just established the African Concessions Syndicate Ltd. (registered on 4 October 1895). This 'syndicate' was awarded the 'preferential right for seventy-five years to generate 250,000 kW at the Falls, and the exclusive right to transmit power from the Falls to the Rand' (Christie, 1984: 28). In 1894, Rhodes had requested and received a positive expert opinion on the feasibility of this project from George Forbes, who had been a consulting engineer to the Niagara power works. The sale of this concession on 14 December 1906 to the VFTPC marked the beginning of one of the largest power schemes to be established in Johannesburg before the First World War.

The African Concessions Syndicate Ltd. was a subsidiary of the British South Africa Company. Extensive rights and privileges were conferred to the Syndicate by its British parent company. The BSAC, in turn, held vast powers that are usually assigned to governmental authorities or state bodies, such as the right to military interventions. This company effectively operated as a 'commercial entity' with sovereign powers (Miles, 2013: 41) and merged functions that are usually carried out separately by the state and investors (Miles, 2013: 33). In fact, the BSAC claimed these rights on the basis of its status as a Chartered Company under the British Crown, which endorsed it to perform a colonial mission in Southern Africa, including the conquest of land, mineral and property rights.

However, on what grounds was the BSAC entitled to concede the rights to harness power from the Victoria Falls (and transport it to the mines of the Witwatersrand) to a subsidiary company, the African Concessions Syndicate? The BSAC claimed these rights to the Victoria Falls territory based on 'the Rudd concession', which was acquired in 1888, the year that the BSAC was established.[188] This concession was acquired to colonize Mashonaland and Matabeleland in Southern Rhodesia. The way in which this concession was acquired[189] is crucial, because the rights conferred in this concession were eventually purchased by the VFTPC, and in fact, the acquisition of these rights was the reason for its existence. The concession consists of a document, signed by the King of the Ndebele, Lobengula, under deceptive conditions. It ceded rights to mine and administer but not to occupy land. Nevertheless, the BSAC in 1890 occupied and started settling Mashonaland. Effectively, the land was annexed by Britain through a chartered company under the control of Cecil Rhodes. The complex and contested legal rights and privileges that lie behind these early concessions, especially concerning land and property, persist to this day. They precede the often-cited land losses through the Native Lands Act of 1913, but much dispossession of land occurred before then.

---

188   These claims were challenged by the Lippert Concession in 1891.

189   'Not surprisingly, when European-concessionhunters came with legal documents for the chiefs to sign, the Africans could not conceivably have grasped the implications of what they were signing, especially in the case of grants pertaining to land rights: Their culture orientation did not encompass private ownership of land, let alone the alienation thereof as a commodity. Never realising that the documents they signed and the money they received alienated their land permanently according to European law and in fact prepared them for colonial domination, the chiefs actually perceived themselves as having the power and authority to grant complete rights of usufruct and still have the last say as for the ownership of the land. [...] Lacking in the comprehension of the powers they were up against. The chiefs when they found that they had been caught up in legal niceties with concessionaires, simply offered to return the monies they had been offered. They discovered, often too late, that the monies they received legally and permanently alienated the concessions areas. (Selolwane, 1980: 85–6).

By the time the Transvaal Power Act (which obliged electricity suppliers to apply to the Power Undertaking Board for a licence) came into force in 1910, the VFTPC had already acquired all of the significant concessions and rights to generate and transmit power in Johannesburg. The Act essentially endorsed the VFTPC's monopoly, with the option to expropriate the company after 35 years. When the Electricity Act established Escom and the Electricity Control Board in 1922, Escom effectively remained outside of government control.[190]

Concessions designate a bundle of rights with varying powers, competences, authorities and jurisdictions. Overall, however, unlike Hughes' case studies in Chicago, Berlin and London, where concessions were issued through the political procedures of stable local or national governments, the concessions in Johannesburg were issued by transitory (and at times precarious) agencies that, in turn, operated within the opaque, contradictory and changing legal and political circumstances of imperially contested territories. Up until 1910, these concessions were issued in the absence of legislation or regulatory frameworks. By this time, however, the formation of a power supply scheme that would determine electricity provision in Johannesburg over the next 40 years had already been completed.

The notion of the 'concession' has been used to designate a variety of arrangements (Veeser, 2013; Fischer, 2014). Concessions in the colonial context were transactions of power on a grand scale. They operated at the intersection of 'the historical narrative of empire and international investment law' (Miles, 2013: 28). Concessions also were used as a tool to play out 'political and commercial rivalries [...] amongst the European capital-exporting states' (Miles, 2013: 30).[191] For example, Selolwane considers the

---

190   The ECB was given regulatory powers to issue licences to anyone supplying electricity except 'local authorities, the South African Railways and Harbours, a government department' or 'self-producers' (Marquard, 2006: 148).

191   Miles considers concession agreements often to have been exploitative: 'The rights obtained by concessionaires were often extensive, involving jurisdictional control of substantial areas of land and significant natural resources for lengthy terms in return for payment of royalties. The scope of individual agreements varied, and, although this type of arrangement often concerned only an isolated enterprise, it still effectively involved

colonization of Bechuanaland to have mainly taken place 'through concession rather than conquest' (Selolwane, 1980: 76). The BSAC acquired a legal doctrine of entitlement 'to enter into treaties, found and administer settlements, engage in military conquest, and build forts' that enabled 'nonsovereign actors to operate in the international sphere' (Miles, 2013: 34). In territories ruled by colonial powers, special privileges and monopoly rights were often added to concessions in order to attract capital from foreign investors (Veeser, 2013: 1136). Selolwane views the colonial activities of 'commercial' companies, such as the BSAC, 'as major agents of capitalist expansion that constituted private rather than public imperialism' (Selolwane, 1980: 76). Concessions in such circumstances often functioned as tools 'to bridge the gap between the legal systems' of states (Veeser, 2013: 1142). Sometimes concessions granted in colonial territories also gave companies 'the right to compel labour' (Veeser, 2013: 1143). Moreover, concession agreements also involved the creation of settlements. This policy was pursued in the case of the power stations in Johannesburg, which were also built with adjacent settlements. Such overseas concessions were 'international legal doctrines [that] were developed and moulded to legitimise the use of oppressive techniques by European powers throughout the colonial encounter' (Miles, 2013: 31). According to Miles, international law 'was an important tool in facilitating the objectives of Western commercial and political hegemony' (Miles, 2013: 31). Categories such as 'civilised' and 'uncivilised' were used to justify the exclusion of non-European communities from international law (Miles, 2013: 31).

Miles observes that concessions were protected and legitimized by 'international rules on investment protection' (Miles, 2013: 28). From the rules that were set up by European nations emerged legal principles that 'became part of the process of building and maintaining Western economic and political dominance'. This process 'evolved into imposed assertions of universally applicable international law as the colonial encounter unfolded' (Miles, 2013: 29). Selolwane proposes that 'concession-acquisition came to play an important role in the expansion of monopoly capitalism', and she

---

the transfer of sovereign rights held by the state to the holder of the concession' (Miles, 2013: 28).

considers concessions to be a tool of imperial conquest (Selolwane, 1980: 79). She views the pursuit of concession hunting 'as a minor agency of capitalist expansion' because '[o]nce acquired, a concession was valueless until and unless it could be taken over by a company with substantial capital backing' (Selolwane, 1980: 79). By granting the BSAC a British royal charter in 1889, 'the British imperial government was merely establishing political control over the process of concession-colonization spearheaded by various companies, and through the cheapest way possible' (Selolwane, 1980: 113). The BSAC was granted extraordinary powers by the United Kingdom 'to enter into treaties, to annex territory, and to obtain commercial concession agreements'.[192] Through these concessions, the BSAC 'acquired wide-ranging trading, investment, and jurisdictional rights' (Miles, 2013: 30). The concession sold by the BSAC to the African Concessions Syndicate to supply power at the Victoria Falls represents a 'transfer of economic and jurisdictional control' that 'contributed to the process of infusing European notions of property rights and the creation of replacement legal regimes' (Miles, 2013: 31).

Whereas Hughes considers concessions to be one of many components in the 'seamless web' of society and technology, the above deliberations draw a different picture. For regions that fell under the competitive imperial contest of European nations, the wide-ranging scope of concessions required for the establishment of electric power stations formed part of an emerging international legal order. It endorsed the interests of imperial powers and served as a legal sanction to accommodate inconsistencies, to justify violence, and to transcend contradictory parallel jurisdictions (Miles, 2013: 33). Concessions were essential for the development of the electric power scheme in Johannesburg, yet in their certified function as legal instruments, they silence the vast scope of issues that were managed under this label. However, these issues are important indicators of the underlying forces that set the terms of debate for electrification in Johannesburg and restricted the range of possible paths to be taken.

---

[192] Charter of the British South Africa Company, London Gazette, 20 December 1889.

Natives

Of all the problems with which the present and future statesmen of South Africa are faced, none can compare in gravity or complexity with the native question. Stated in brief terms, it may be defined as a standard of the relations between the white man and the black man.
Lionel Phillips, 1905.
On 'The Native Problem'

What I would like in regard to a native area is that there should be no white men in its midst. I hold that the natives should be apart from white men, and not mixed up with them.
Now, I say the natives are children. They are just emerging from barbarism. They have human minds, and I would like them to devote themselves wholly to the local matters that surround them and appeal to them.
Cecil Rhodes: His Political Life and Speeches, 1881–1900.

Rough guesses place the number of natives at from two to ten millions, but, as a matter of fact, no one knows even approximately their number. This lack of information is due to the roving propensities of the natives. Here to-day, there to-morrow, it would take a mightier hunter than even the famed Selous[193] to hunt them all down.
Edgar Mels, *Scientific American*, 1900.

The small lighting systems of the 1880s evolved into the regional power systems of Johannesburg and the Rand against the background of migrating populations before the First World War. These included indigenous populations, people from adjacent colonies (e.g. Portuguese), European and American immigrants, and Boer and English settlers. From the very beginning, these populations were organized into a hierarchical labour administration that was based on the idea of racial supremacy over African populations. The labour policy of the gold mines followed three main categories to classify the work and wages in the mines, ranging from highly skilled

---

193   The British explorer and hunter Frederick Courteney Selous (1851–1917) was active in Southern Africa.

mining engineers to unskilled workers: indigenous peoples of Southern Africa for 'unskilled labour'; imported European and American skilled labour;
and local 'White' skilled labour. In 1897, at the time the GEPC was constructed, the chief engineer of the Consolidated Gold Fields Group said that
the workers at the Witwatersrand mines comprised more than 9,000 'white
employees' with total annual wages over $9,000,000, and 70,000 'Kafirs', with
total annual wages of nearly $12,500,000 (Hammond, 1897: 240).

This socio-political and administrative order shaped the particular
course of the history of electrification in Johannesburg. Conversely, the
maintenance and consolidation of this oppressive administration was supported by the institutions of the electric power industry. The VFTPC also
assigned its employees to three categories: 'professionals (well-paid and
dependable), white artisans and supervisors (not so well-paid, far less
dependable), and black workers (poorly paid, housed in compounds for control, and substitutable)' (Christie, 1984: 57).

The category of 'the native' stands at the centre of this classification. He
typically appears in connection with the labour requirements of the gold
mining industry. The gold mines regularly lamented the shortage of unskilled labour, but at the same time, they consistently endeavoured to cut the
costs for this type of labour. These circumstances favoured the mines' decisions to mechanize labour through new technology and power sources. The
first lights and generators in Johannesburg were installed at the gold mines.
The first two power stations in Johannesburg came into operation at a time
when the surface outcrops started hitting lower levels that commanded new
deep-level mining procedures. These new mining practices compelled the
gold mines to reach a new balance between human labour and power for
machinery. Thus, the amount of electric power supplied to foreign-owned
gold mines in the late 1890s was directly related to labour costs, 'the crucial
area of cost minimisation' in the gold mining industry (Richardson & Van
Helten, 1982: 81).[194] The entitlement of foreign companies to mining and

---

[194] 'In 1898, for example, 58 producing companies spent 53–44 per cent of their total
production costs on labour, and the balance on various stores and fuel, the largest single
items being explosives (10.95 per cent) and fuel (8.23 per cent). Consequently, commodity price levels and the costs of different types of labour-power were a central concern of

exploration rights entailed an associated sense of entitlement to 'native' labour.[195]

Therefore, the production process adapted to save costs, 'exploited the growing reservoirs of cheap unskilled African labour' (Richardson & Van Helten, 1982: 81) and 'the ratio between European and African (and later Chinese) miners, both in terms of numbers and relative costs, became one of the most sensitive indices of the profitability of mining operations on the Rand' (Richardson & Van Helten, 1982: 81). This ratio fuelled the rhetoric about the expansion of the mines to argue that 'expansion depended to a very large extent upon the movement of this ratio in favour of African labour' (Richardson & Van Helten, 1982: 81).

Because of the 'problem' of the 'native labour supply' for the gold mines, the Transvaal Chamber of Mines experimented with various plans to recruit unskilled workers. Such plans were supported by several legal regulations that aimed to control the movement and residence of 'natives' in Southern Africa, and they endorsed a racial division of labour by imposing a 'colour bar'.[196] Examples of these regulations include the Transvaal Native Pass Law of 1895[197] (drafted by the Chamber of Mines and adopted by the Transvaal

---

the industry's managers. As the industry was not generally able to exercise a controlling influence over commodity prices, labour costs thus became the crucial area of cost minimisation' (Richardson & Van Helten, 1982: 81).

195 'The technical difficulties of deep-level mining, the scale of investment that it required, and the absence of an indigenous skilled workforce meant that in the initial stages of capital accumulation the industry was forced to resort to the introduction of skilled immigrant workers to perform certain specific tasks of production and to oversee the production process in general. However, the reliance upon the relatively scarce and therefore expensive skills of immigrant miners from Europe, North America and Australasia to perform these tasks was in direct conflict with the cost-minimising strategies dictated by the imperatives of profitable production' (Richardson & Van Helten, 1982: 81).

196 These policies continued legislation that had previously been passed to regulate the 'native' population in Southern Africa, such as the *Masters and Servants Ordinance* (passed to deprive black tenants of legal protection by defining them as 'servants' instead of wage labourers) or the Glen Grey Act (passed by the Cape Parliament by Cecil Rhodes in 1894).

197 The Pass Law No. 31, of 1896, amendment of Law No. 23 of 1895: '[…] so it is hereby enacted as follows: Regulations in terms of Article 88 of the Gold Law. For the purpose of facilitating and promoting the supply of native labour on the public diggings of this

government to incorporate Africans into wage labour),[198] the Liquor Law of 1896 and the Transvaal Labour Importation Ordinance of 1904 (designed to enable the importation of Chinese labourers).

In 1897, an Industrial Commission of Inquiry was convened to examine the problems of the gold mining industry. One of the main obstacles identified was the inability to find an 'adequate supply of native labour at reasonable pay' (Chamber of Mines, 1897: iii).[199] The Commission's report included a detailed overview of 'native wages' (see Figure 8).

One of the means of controlling and regulating the native labour force was the Pass Law.[200] The 'native' was obliged to register and apply for a pass

Republic, and for the better controlling and regulating of the natives employed, and the relations of employer and native labourer' (Industrial Commission of Inquiry, 1897: 576).

198 'Regarding the Pass Law, there has been, as far as I am aware, no witness yet before the Commission who has stated that this law, as administered, had benefited his company, and Mr. Goldmann has informed you that out of thirty-three companies employing 19,000 boys monthly, 14,000 have deserted since the new Pass Law came into operation, without any single one of these deserters having been brought back to the mines and justice. In my opinion, the Pass Law, though good as a temporary expedient, is only the kindergarten of the native question [...]' (Mr Jenning's Evidence, Industrial Commission of Inquiry, 1897: 219).

199 'Now, regarding the native labour, which comprises in numbers by far the greater proportion of labour we are using. What has been the keynote of our trouble? Lack of supply in proportion to the demand, and inefficiency and ignorance of this class of labour which is not trained to the intricate work demanded of it. Far more skill is required of the kaffir on the Witwatersrand than is the case for the most part on the diamond fields, or in any agricultural pursuit in South Africa. The boys come here raw, some very young, often with weak physique, and are all comprised in the same classification. They are accustomed to their own simply ways, and desire to return to them as soon as possible. They come, in fact, only in order to make enough money to return to their kraals with sufficient means to enable them to marry and live in indolence. There is much latent possibility in them for learning, but they leave us often as soon as they become really useful, and by the various companies vieing with each other to obtain their services, they have become masters of the labour situation' (Mr Jennings's Evidence, Industrial Commission of Inquiry, 1897: 219).

200 'The whole object of the Pass Law is the control and regulation of native labour on the goldfields, and if the desertion of natives can be prevented, the mining industry

XIV,

SCHEDULE OF NATIVE WAGES, WITWATERSRAND.

Agreed to at a combined meeting of Mining Companies, May, 1897.

MINE.		s.	d.	MILL.		s.	d.
Machine helpers	... ...	1	8	Elevator boys	... ...	2	0
Hammer boys	... ...	1	6	Vanner boys	... ...	2	0
Shovel boys	...	1	3	Mill boys (12 hours)	...	2	0
Tram boys (10 feet trucks)	...	1	2	Mill boys (eight hours)	...	1	4
Tram boys (16 feet trucks)	...	1	6	Blanket and sluice boys	...	2	0
Dry shaft and winze boys	...	1	8	Crusher boys	... ...	1	4
Wet shaft boys	... ...	2	0	Surface trammers	... ...	1	9
Wet shaft boys (when develop-				Mule drivers	... ...	2	6
ing only)	... ...	2	6				
Boys cutting hitches for timber		1	6	CYANIDE.			
Timber boys	... ...	1	2				
Stope gangers' assistants	...	2	0	Solution shed boys ...	...	1	4
Station boys (where white man				Boys (filling and discharging)		1	9
employed)	... ...	1	2	Zinc cutters	... ...	1	6
Station boys (where no white				Tramming residues...	...	1	4
man employed)	... ...	2	6				
Air hoist drivers	... ...	2	0	GENERAL.			
Pumpman's assistants	...	1	8				
Platelayer's assistants	...	1	6	Fitters' boys	... ...	1	6
Pipeman's assistants	...	1	6	Blacksmiths' boys (strikers) ...		2	6
				Blacksmiths' boys (helpers)	...	1	4
SURFACE.				Carpenters' boys	... ...	1	2
				Masons' boys	... ...	1	4
Stokers (12 hours) ...	...	2	6	Police	... ...	2	6
Stokers (eight hours)	...	1	8	Compound cooks	... ...	2	0
Engine cleaners	... ...	1	6	Drill packers	... ...	1	0
Sorting boys	... ...	2	0	Drill sorters	... ...	1	6
Head-gear boys (where white				Surface labourers	... ...	1	2
man employed) ...	...	1	4	Office and store boys	...	2	6
Head-gear boys (where no white				Assay office boys	... ...	2	6
man employed) ..	...	2	6	Coal boys (off-loading)	...	1	6

NOTES.—Timber boys assisting in timbering shafts, to be paid at the rate of wet and dry shaft boys.

Seven and one-half per cent. of the natives employed may be paid special rates.

Month to be reckoned as consisting of at least 30 working days.

Figure 8: Schedule of Native Wages, Witwatersrand. Agreed to at a combined meeting of Mining Companies, May 1897 (Industrial Commission of Inquiry, 1897).

would be fostered [...]' (Mr F.W. Kock's Evidence, Industrial Commission of Inquiry, 1897: 296/297).

and badge in order to be identified by pass officials.[201] The following paragraphs provide excerpts from the Pass Law No. 31, 1896 to illustrate the detailed administration that was established to deal with 'the native'.

6. The Mining Commissioner or pass officer appointed for the district shall enter in a register to be kept for the purpose, of the form of schedule B hereto, the name of the native, his tribe, chief, father, district or country, stature and marks, if any, etc., and he shall also number each native consecutively. Such registered number shall thereafter be the native's official mark so long as he remains within the district.

7. In addition to the district pass to form A, the pass officer shall at the same time issue to each native a metal ticket or badge, on which shall be clearly stamped or impressed at the time of issue the native's registered number, the initial letters of the labour district, and year of issue. This ticket or badge shall be attached to a strong leather strap or buckle, and must be worn by the native round his left arm about the elbow, so as always to be clearly visible. Such district pass and badge shall be issued free of charge.

8. Such district pass and metal badge shall enable and authorize the native to whom it is issued, to seek employment within the labour district for which it is issued for a period of three days from the date of issue.

9. If the native fails to find employment within the prescribed three days from the date of issue of district pass, or after discharge by his last employer, he shall return to the Mining Commissioner or pass officer who issued the pass and badge, and may have an extension of a further three days endorsed thereon by the pass officer, on payment of a fee of two shillings [...].

11. A native working on a proclaimed goldfield, and wishing to remove from one labour district to another, or such or any other proclaimed goldfields, shall first

---

201 'It thus speaks for itself that, perhaps with a few exceptions, deserting natives are punished. It has to be acknowledged that when a native throws away his passes and badge, he cannot again easily be identified by any of the pass officials, and I challenge anyone to describe a native, and register him in such a way that he would be able to identify him without the aid of his passes and badge, out of 60,000 other natives [...]. Identification would, therefore, be easier by the officials of the different mines, and in order to meet this I have suggested a charge office, with a large yard, where all natives arrested could be kept for a certain time to ensure identification by their employers. According to my opinion, the question is not so much the detection and apprehension of deserters, but the prevention of desertion [...] this may be accomplished in two ways [...].(2) By making the punishment for desertion so severe that no native will venture to leave his employer illegally [...]' (Mr F.W. Kock's Evidence, Industrial Commission of Inquiry, 1897: 297).

apply for leave to do so from the Mining Commissioner or other appointed pass officer in his district, and such leave shall be granted him, provided he then holds a district pass, in clear order, with metal badge, and that his last employer, if any, shall have filled in the full discharge required on the district pass, form A [...].

14. Any native found in the labour district without the distinct pass of form A and metal badge, or without a travelling pass, or any defaulter under Article 9, shall be punishable by a fine of not exceeding (pounds) 3, or not more than three weeks' imprisonment with hard labour for first offence, and for second offence a fine not exceeding (pounds) 5, or not more than four weeks' imprisonment with hard labour and lashes, and at the discretion of the Court before whom he shall be convicted for every offence thereafter.

17. The Government shall make arrangements as it may deem desirable, so that each labour district in a proclaimed goldfield shall use and issue a pass distinctive in colour from those of the other labour districts, and metal badges with initial letters of each district, in order to facilitate the detection of vagrants or natives moving in any labour district without a pass and badge for that district.

Employers with more than twenty native labourers are obliged to keep a register according to schedule form F, at the end of each month fill in and return to Mining Commissioner or pass officer of district, with all details of natives engaged and discharged that month.

27. Government shall appoint such officers in each district as may from time to time be found necessary for the due and proper administration of these regulations, and shall further appoint special labour inspectors for each labour district with power to summarily arrest all natives contravening these regulations. The duties of such special labour inspectors shall inter alia be

(a) to make regular and frequent inspections of registers of native labour kept by employers, to inspect all natives employed, to ensure that their badges are worn and in order, and, if need be, compare any or all such natives with the district pass filed by the employer (Industrial Commission of Inquiry, 1897: 577–8).[202]

The category of the 'native', and the rules set out to regulate his whereabouts and civil rights, were not only used for economic purposes. It served the imperial objective to appropriate, settle and civilize the territories of

---

[202] Extracts from the Pass Law No. 31, 1896 concerning the administration of 'natives' (*The Mining Industry. Evidence and Report of the Industrial Commission of Inquiry.* Compiled and published by the Witwatersrand Chamber of Mines, Johannesburg, 1897).

Southern Africa.[203] Accordingly, the labour scheme was not only developed to provide manual labour. It served to appropriate and civilize the country by white settlers. In this sense, the idea of the 'native' also provides a foundational condition for the establishment of the first two power stations in Johannesburg, which were owned by a German bank and a British gold mining house. German imperial policy in Southern Africa primarily pursued economic and industrial rather than political colonization. A quote from Walther Rathenau provides the key to this sense of entitlement: 'If the nigger possessed the characteristics of the European, we would have no right to colonize his country' (Rathenau, Reflexionen 1908 [2008]: 163, my translation).[204] Rathenau's quote indicates that the idea of a subordinate race provided imperial Europe with a simple justification to take away the land and rights to self-government of the local populations. The idea legitimized the subjugation of people, which was required to dispossess and relocate them, and it allowed the Europeans to ignore them in their plans for the industrial development of Southern African territories – save in their capacity as labourers.

Who was subsumed under the category of 'the native'? An article in the journal *Scientific American* in 1900 on the 'The Natives of South Africa' offers the following description.

> The writer has seen the South African native, commonly called Kafir, in all his varying phases, in his wild state, semi-civilized and wholly so. He has seen the native at his best and at his worst – untainted by the touch of civilization and soiled by its proximity. And through it all, the writer has believed, and perhaps always will, that the Kafir, whether Zulu or Basuto or Becbuana or Swazie or Amatonga or Matabele or any other tribe, has good in him, just as though his skin were white – and bad too. Summed up in a few words, the Kafir, in his uncivilized state, is an overgrown child, with childish foibles and shortcomings. But let him learn the vices of civilization, let him realize the evil there is in him, let him discover that there is a broad

---

203 For example, mining magnate Lionel Phillips (1855–1936), friend of Cecil Rhodes and Alfred Beit, stated, 'It will be long before the natives derive the full benefit of closer contact with civilisation, because their reasoning powers are limited, and their point of view entirely different to ours' (Phillips, 1905: 99).

204 'Besässe der Neger die Eigenschaften des Europäers, so hätten wir kein Recht, sein Land zu kolonisieren' (Rathenau, Reflexionen 1908 [2008]: 163).

path leading to destruction – and you will find a fully civilized being, as capable in
certain directions as is the white man (Mels, 1900: 56).

In 1897, the Industrial Commission of Inquiry defined the category of the
native to 'apply to males of all the native and coloured races of South Africa'
(Industrial Commission of Inquiry, 1897: 576). This category was not just
used in juxtaposition to the 'white' population; it also related to a third category, that of the 'coloured person', which had been defined in the Gold Law
of 1896 to 'signify every African, Asiatic native, or coloured American person, Coolie or Chinese' (Industrial Commission of Inquiry, 1897: 599).
These categories, however, in no way reflected the populations in Johannesburg, and difficulties emerged in practice.

> Then the difficulty arises with the people known as Cape boys and bastards [...].
> But you have got the question of white people, you have got the mixed breed, and
> you have got the pure kaffir. [...] As illustrating the difficulty, I will give you a case
> which happened to myself. I have a man in my employ and he is very dark, but his
> father is an Englishman, who is married to a woman who is descended from a St.
> Helena woman and a white man. Out of nine children this man is the dark one; all
> the others are white (Mr. F.W. Kock's Evidence Industrial Commission of Inquiry,
> 1897: 298–9).

The categories sanctioned the idea of the white settlement of Southern African territories: civilization. This entitlement to civilize the 'native' was shared
by many powerful personalities involved in the business of electrification in
the Transvaal, such as Emil Rathenau, Cecil Rhodes, Alfred Beit and John
Hays Hammond. Cecil Rhodes probably best illustrates the sense of entitlement over 'natives' that gave him free reign to develop his plans to develop a
British colonial empire from the Cape to Cairo. The establishment of the
central power station to supply mines of the Consolidated Gold Fields group
in the Transvaal was only the first step in his grand plan. At the time, the
independent Boer Republic in the Transvaal stood in the way of his creating
a South African union under the British flag. As early as 1894, Rhodes had
consulted with George Forbes (consulting engineer to the Niagara Falls power

project) about the possibility of generating electric power at the Victoria Falls and transmitting it to the gold mines of the Witwatersrand.²⁰⁵ Walther Rathenau, one of the masterminds behind the first regional power scheme in Johannesburg, used the expression 'nigger' to discuss the 'native' question.

> The nigger differs mentally from Occidentals through his greatly reduced capacity for abstraction and concentration. General and ideational concepts are virtually inconceivable to his thinking, which is not unskilled in manual work; to retain an interest and thinking all the way through to a final result causes him pains, he evades it [...]. For this reason a well-founded mental development of the nigger will remain wishful thinking in the foreseeable future (Rathenau, 1908 [2008]: 157; my translation).²⁰⁶

---

205 'In 1894, before I had completed the first electric works at Niagara, I was asked by letter from Johannesburg whether it would be possible to transmit power from the Victoria Falls, on the Zambesi to all the gold mines in Rhodesia, varying from 350 to 500 miles distance. At first I was inclined to throw the letter into the waste-paper basket. No one, up to that date, had, to my knowledge, seriously considered the financial aspects of so distant a transmission of electric power. But the letter required an answer; so I sat down to work out and quote some figures which should show the absurdity of the scheme. I had been supplied with maps and costs of fuel, etc. and in a short time I found, to my astonishment, that if the facts were as stated, the scheme was financially and electrically a sound one [...]. Upon this I was asked to go to South Africa to negotiate with Mr. Cecil Rhodes and Dr. Jameson for a concession. They both appreciated the value of the enterprise to the country, and prepared a draft of the concession, which was only awaiting the sanction of the Chartered Company's Board when the Jameson raid and the Matabele rising closed negotiations. Here is a case where, if there be really good gold mines, it will pay handsomely to transmit electric energy a distance of 500 miles, provided the surveys of the Falls prove as satisfactory as the photographs do, and provided the fever is not an insurmountable obstacle' (Forbes, 1898: 27).

206 'Der Neger unterscheidet sich geistig vom Okzidentalen durch weit herabgesetzte Fähigkeit zur Abstraktion und Konzentration. Allgemeine und ideelle Begriffe sind seinem im Handgreiflichen nicht ungewandten Denken nahezu unfassbar; andauerndes, bis zum Endergebnis nachgehaltenes Interesse und Nachdenken bereiten ihm Schmerzen, er weicht ihm aus. [...] Deshalb wird eine festgegründete geistige Entwicklung des Negers für absehbare Zeit ein frommer Wunsch bleiben [...]' (Rathenau, 1908 [2008]: 157).

Although Walther Rathenau conceded that 'a certain legal consciousness' and 'sense of justice' formed part of the few abstractions the 'nigger' was capable of,[207] he concluded that education would remain of little importance to African economic development, except for training in manual skills.[208] The particular geography of imagination of this influential historical figure for the histories of electrification in Germany and South Africa is significant: Walther Rathenau was the son of Emil Rathenau, who had purchased the Edison Patents for Germany and founded the A.E.G. He was a director of the A.E.G., he headed its central power station division before Klingenberg took over in 1899, and he was the director of the Berliner Handelsgesellschaft, which provided finances for the A.E.G. and (with the Deutsche Bank) for the establishment of A. Goerz GmbH in 1892, which had commissioned Johannesburg's first power station.

A few examples of the era's prevalent images of the 'native' are worth quoting to illustrate how this category was presented. The American mining engineer for the British South Africa Company, John Hays Hammond, one of Rhodes' close friends,[209] writing on the topic of 'South Africa and its Future', described the populations of South Africa as follows:

> South Africa has a heterogeneous population of about five million, of which over six hundred and fifty thousand are whites, English and Dutch preponderating [...]. It is to the unique colonizing capacity of Great Britain that South Africa owes the important position that she to-day holds. Of the native population the bulk belongs to the Bantu family, which occupies all of Central and South Africa, and forms the great reservoir from which the manual labor of South Africa is drawn. The members of this family are generally designated 'Natives' or 'Kafirs'. [...] Kafirs are the negroes of South Africa, though they have characteristic differences from the negroes of the West Coast of Africa, whence came our American negroes. The Bantu tribes are not

---

207   ,Zu den wenigen Abstraktionen, deren der Neger fähig ist, gehört ein gewisses Rechtsbewusstsein und ein deutlicher Gerechtigkeitssinn [...] (Rathenau, 2008: 163).

208   ,Erziehung wird daher, soweit sie nicht auf Erlernung einzelner Fertigkeiten, Notionen und Handgriffe hinausläuft, sondern ihren idealen Weg als Geisteskultivation verfolgt, ein für die afrikanische Wirtschaftsentwicklung wenig bedeutender Faktor bleiben' (Rathenau, 1908 [2008]: 157).

209   '"Rhodes," said Mr. Hammond, "was by far the greatest man I have ever met. He had unlimited vision, extraordinary perception, unbounded courage."' (Forbes, 1917: 187).

aborigines, having come from Northern Africa. It is a remarkably prolific race, the numbers of which are increasing with great rapidity. Of far less importance numerically are the Bushmen, the true autochthons, but now the social pariahs of the whole continent. The interior of South Africa was originally sparsely populated by these pigmy Bushmen tribes, which have resisted all attempts at civilization, preserving their nomadic instincts, and still approximating to the lowest known species of humanity. The Hottentots, another tribe, which had located in South Africa interior at the advent of the whites, though possessing some resemblance to the Bushmen, have radical ethnological differences [...]. The Hottentots, though evincing more receptivity, have been but inconsiderably affected by the civilizing influence of their environment, and may with the Bushmen be set down as unimportant, and indeed rapidly vanishing, factors in South Africa's future (Hammond, 1897: 234–5).

Hammond's views illustrate the classification of the 'native' as a source of labour rather than as citizen.[210] In his imaginary geography, the 'native' was viewed as a labourer rather than a citizen, as described in Cecil Rhodes' second reading of the Glen Grey Act in the Cape House on 30 July 1894, speaking on 'The Native Question'.

What I have found is this, that we must give some gentle stimulus to these people to make them go on working. There are a large number of young men in these locations who are like younger sons at home [...]. These young natives live in the native areas and locations with their fathers and mothers, and never do one stroke of work. But if a labour tax of 10 s were imposed, they would have to work [...]. We want to get hold of these young men and make them go out to work, and the only way to do this is to compel them to pay a certain labour tax (Rhodes, in Verschoyle, 1900: 381).

The connection between the history of electrification and oppression by the non-native population is not apparent and cannot be easily traced. The geography of imagination of key players in the history of electrification of the regional system on the Rand nevertheless reveals glimpses of this connection.

---

210 'It cannot be denied that the regime of the white man has greatly ameliorated the condition of the Kafir, who, before his advent, was the victim of internecine wars, and of the operation of a despotic form of native government, possible only among barbaric people. The only alternative supply of labor would be the importation of indentured East Indians [...]' (Hammond, 1897: 247).

Despite their selective and non-representative status, these glimpses are important historical markers. Thomas Edison's 'ordinary native', Walther Rathenau's 'nigger' and Charles Merz's 'White population' all map this imaginary geography. These categories were employed to assign the subordinate status to indigenous populations. Hughes' history of electrification of Western society, despite its focus on three imperial nations at the height of a global imperial scramble for territory, does not record the presence of such populations. The subordinate status and invisibility of these populations are effectively endorsed by the application of Hughes' historical model of technological development.

The technological projects of the RCEW and the GEPC, rather than expressing a response to local needs for electricity, served to foster the interests of rival imperial powers and their companies' offspring in Johannesburg. The provision of electricity to the city of Johannesburg was an intermittent, insignificant by-product of this imperial enterprise. The subsequent history of the electrification of Johannesburg and South Africa bears this imprint. By 1898, the South African Republic (Transvaal) had become the world's largest producer of gold, accounting for nearly 28 % of output (Tuffnell, 2015: 56). The first two central power stations were both associated with companies and business ventures involved in this grand imperial exploration project. Both the RCEW and the GEPC were private investments. These investors pursued the long-term economic and political imperial interests of Great Britain and Germany in the Transvaal. Thus, the power stations represent two competing imperial nations in Johannesburg during the years leading up to the second Boer War. Rather than simple investment or project development enterprises, their geography of management reflects the intricate political and economic scramble for territorial supremacy in the age of New Imperialism. This scramble rested upon a sense of entitlement to operate and intervene in foreign territories. Company ventures were one of the principal vehicles that foreign nations used to participate competitively in this scramble.

The above discussion suggests three common attributes that find their expression in the recurring themes that have been identified in the history of electrification in Johannesburg: gold mining, empire, national competition, companies, concessions and natives. These attributes relate to the European

'Zeitgeist' and describe forces that powered the course of this history: a *sense of entitlement*, a *sense of appropriation* and a *sense of intellectual ascendancy*.

Hughes in Johannesburg: New Concepts

The cultural forces influencing the systems stemmed from the societies within which the systems grew. These societies were of various kinds according to time and place.
Thomas P. Hughes, 1983.

This chapter presents several new concepts to describe the dynamics and drivers of electrification; the *frontineer*, the *settler company*, *technological entitlement*, *technological scramble*, *global texploration* and *texpansion*. These concepts are introduced by way of examples that relate the early history of electrification in Johannesburg to Hughes' history of electrification and model of technological change.

Hughes' book uses two points of entry to study the history of electrification. On the one hand, he pursues the material expansion of technology relating to electric light and central power stations. On the other hand, he traces 'the decisions made by inventors, engineers, managers, and financiers who were system builders' (Hughes, 1983: x). According to Hughes, the focus on system builders also supports his decision to undertake a comparative study of the interaction between region and technology rather than between nation and technology. Whereas the material evolution of technology may be drawn from historical records and leaves limited scope for historical controversy, the choice of system builders is a problem of methodology in need of explicit criteria for selection. Hughes' system builders are defined by professional categories, and they include inventors, engineers, managers and financiers. All of these professions (individually and in any combination) are capable of presiding over a system-building process. Hughes' criteria for the selection of system builders, therefore, pertains to their professional qualities and distinctions, which, in turn, are defined by their material utility in assembling the technology of electric power systems.

System building, according to Hughes, involves solving problems of various kinds at various stages in the history of technology. In Hughes' framework, system builders may be discerned retrospectively by way of their problem-solving capacity. Put differently, a solved technological problem will lead the researcher to at least one system builder. In this way, Hughes constructs his theoretical framework of growth, which is driven by the reverse salient and critical problems.[211] Thus, the historian's task is to identify the key problems that were solved in the history of electrification and to trace the system builders who solved them in their specific cultural setting. Electrification remains equated with the setting up of electric lights and the construction of regional power systems. However, Hughes' notion of the system builder also uses undeclared selection criteria. The system builders in Hughes' history of electrification are all American, German and British citizens. Hughes deduces a model of technological change from case studies in these three countries and extrapolates it on a global scale.

System builders achieve system goals by correcting reverse salients. The system goal is a compelling technological condition that provides guidance and direction to the expansive aspirations of the system builder.[212] Hughes' system is composed of related components that are connected by a centrally controlled network. The limits of this control define the limits of the system (Hughes, 1983: 5). Therefore, a key question concerns how the researcher is to ascertain these limits. The study of Johannesburg has shown that this territory was within the limits of control of the key protagonists in Hughes'

---

211   'Innumerable (probably most) inventions and technological developments result from efforts to correct reverse salients. Outstanding inventions and developments in electric lighting and power during the two decades after 1880 were responses to reverse salients' (Hughes, 1983: 80).

212   'The idea of a reverse salient suggests the need for concentrated action (invention and development) if expansion is to proceed. A reverse salient appears in an expanding system when a component of the system does not march along harmoniously with other components. As the system evolves toward a goal, some components fall behind or out of line. As a result of the reverse salient, growth of the entire enterprise is hampered, or thwarted, and thus remedial action is required. The reverse salient usually appears as a result of accidents and confluences that persons presiding over or managing the system do not foresee, or, if they do foresee them, are unable to counter expeditiously. […] The reverse salient will not be seen, however, unless inventors, engineers, and others view the technology as a goal-seeking system' (Hughes, 1983: 79–80).

history, such as Edison, Klingenberg, Rathenau, Mershon and Merz. By implication, there is no rational basis for placing Johannesburg outside of Hughes' system of electrification in Western society; on the contrary, using Hughes' own logic, Johannesburg forms part of his network of interconnected components. This has far-reaching implications, which can be illustrated by Hughes' model system builder, Thomas Edison.

Hughes describes Edison as an inventor-entrepreneur, a 'holistic conceptualizer and determined solver of the problems associated with the growth of systems' (Hughes, 1983: 18): 'Edison focused on one level of the process of technological change – invention – but in order to relate everything to a single, central vision, he had to reach out beyond his special competence to research, develop, finance, and manage his inventions. Because of this organizational, system-building drive, he is known as an inventor-entrepreneur' (Hughes, 1983: 18). Thus, Hughes regards 'the history of Edison systems building' as 'a history of ideas and a study of problem solving': 'Edison's concepts grew out of his need to find organizing principles that were powerful enough to integrate and give purposeful direction to diverse factors and components. The problems emerged as he strove to fulfill his ultimate vision' (Hughes, 1983: 18). Accordingly, Hughes introduces Thomas Edison as the 'Hedgehog'[213] who presided over the process of technological change from problem identification to innovation and technology transfer (Hughes, 1983).

As demonstrated in the study of electrification in Johannesburg, from the very beginning, Edison envisioned a global empire for this technology. Edison prepared the legal and diplomatic conditions to develop this global empire in his agreements with Drexel, Morgan & Co., the company that had also arranged for the financing of his electric light and central power station projects. Edison operated at a global scale through companies and patent

---

213 'Quoting the Greek poet Archilochus, Isaiah Berlin wrote in *The Hedgehog and the Fox:* "The fox knows many things, but the hedgehog knows one big thing." Hedgehogs, according to Berlin, are those "who relate everything to a single central vision, one system less or more coherent or articulate." Foxes, in contrast, pursue many ends, ends that are "often unrelated and even contradictory." Berlin counted Dante, Plato, Lucretius, Pascal, Hegel, Dostoyevsky, Nietzsche, Ibsen, and Proust among the hedgehogs' (Hughes, 1983: 18).

rights for far-away regions before his technology had been developed. The global scope of his project was strategically pursued by skilfully inserting his idea into the European imperial architecture of the late 19th century, with Great Britain as the prime imperial and industrial power, and London as its financial centre. This vision of entitlement to certain global rights, privileges and profits regarding technology was a key driver for these technological projects.

The global dimension and the sphere of control of Edison the system builder challenge Hughes' notion of system limits. More importantly, however, they defy the analytical significance of Hughes' concept of the system builder. Because the system builder is also the solver of critical problems that stem from reverse salients, these concepts, too, need to be revisited. This requires a shift in perspective on the kinds of problems in need of solutions in the history of electrification. Hughes views the problems that Edison set out to solve as reverse salients of material technology, motors of technological growth. The study of Johannesburg suggests that the problem that Edison set out to solve was how to develop technology that would impart to him certain global rights, privileges and profits. One of the consequences of this shift in perspective is that it brings to light a different profile of the performer – one that reaches beyond Hughes' professionalized group of system builders (inventors, engineers, managers and financiers) – and thus suggests a change in the unit of analysis for historical research on technological change.

Key personalities who drove and shaped the history of electrification in Johannesburg between 1880 and 1930 include Cecil Rhodes, Adolf Goerz, Alfred Beit, John Hays Hammond, Emil and Walther Rathenau, Georg Klingenberg, Charles Merz and Hendrik van der Bijl, among others. What key characteristics stand out in these men that allowed them to exert such a powerful influence on the development of electricity? All of these men possessed a frontier mentality, and they all pursued frontier visions. They pushed the frontiers of gold mining, British and German territorial and economic imperial expansion, sales markets for electrical equipment, global investment practices, global company and investment law, 'civilization', mining technology and the associated frontiers of 'native' labour exploitation, empire and monopoly, and of international patent rights and privileges.

Rather than system builders, these personalities may be characterized as *'frontineers'*, a designation that blends the words 'frontier' and 'engineer'. The word 'frontier' refers to two kinds of territorial limits: the frontier of land and the frontier of knowledge. The frontier of land designates the territorial limits of place, whereas the frontier of knowledge refers to the territorial limits of epistemic space. Given its twofold label, the frontineer can be traced methodologically using Trouillot's geographies of management and imagination. Both the frontiers of place and epistemic space hold the promise of new fields for exploitative activity. The frontiers offer playing fields for exploration, expansion and settlement, and they potentially increase spheres of control and influence. The word 'engineer' can be used as a noun or a verb. Hughes uses this category to refer to an occupational activity that was professionalized in parallel with the history of electrification. He coined the phrase 'inventor-engineer' to illustrate that the qualities required to spur technological growth go beyond what has come to be associated with the professional 'engineer' to include inventive, managerial and financial capacities. As a verb, the word is also used more loosely to indicate clever and often secret measures taken to make something happen, something that is to the advantage of the actor. This meaning relates to the etymological roots of the word 'engineer', which derives from the Latin word *ingenerare*, which means to contrive, devise, or create. The dual aspiration to the most advanced territories in place and space is fuelled by a sense of entitlement, a sense of appropriation and a sense of intellectual ascendancy – the foundational drivers of electrification.

Georg Klingenberg was a frontineer: he devised the large electric power system of Johannesburg. He also used this project as a technological prototype, a case study upon which he constructed his theory and wrote his legendary textbook on large electric power stations – defining nothing less than the frontiers of electrical engineering on the eve of the First World War. Simultaneously, however, he also extended the frontiers of his company, the A.E.G., its frontiers of collaboration with the Deutsche Bank and other German and Swiss financial institutions, and the frontiers of the German industrial empire in the Transvaal. Another frontineer, Emil Rathenau, purchased the Edison patent for the electric light not only for Germany but also to establish an industrial empire across the world, epitomized by his trip to America to negotiate with the American General Electric Company for the frontiers of their respective industrial territories. His son, Walther Rathenau, accompanied Bernard Dernburg, the German Secretary for Colonial Affairs, on his travels through Southern Africa and justified German imperialism with the need to expand the frontiers of civilization. The consulting company of the British engineer Charles Merz established offices in London to gain access to the frontiers of electrification in the British Empire.

The concept of the frontineer can be applied inversely as a unit of study to Hughes' history of electrification in Chicago, Berlin and London. The frontineer's properties are valid for the key protagonists of Hughes' history – the 'system builders' Thomas Edison, Samuel Insull, Georg Klingenberg, Oskar von Miller, Charles Merz, Emil Rathenau, George Westinghouse and Werner von Siemens. However, the properties of the frontineer might also make some of these men more prominent than others, or might even bring new actors into focus. In any event, calling these personalities frontineers rather than system builders not only expands the range of potentially qualifying agents but also accentuates their expansionary aspirations and spheres of influence beyond the North Atlantic. For example, the overseas activities and involvements of the Americans Thomas A. Edison and J.P. Morgan, the Germans Emil Rathenau and Georg Klingenberg, and the Britons George Forbes and Charles Merz influenced the technological models, companies and ideas that they applied at home.

The frontineer in the history of electrification blends a sense of entitlement with a sense of appropriation and intellectual ascendancy. These properties may be captured by the notion of 'technological entitlement'. This notion designates a compulsory property of the frontineer and guides his actions for technological change. It refers to the conviction of having an intrinsic right to possess, know or do something relating to technology. Technological entitlement marks out certain privileges of the frontineer of electric power schemes. The frontineer needs this property to be part of the process of electrification.

Hughes' model of system growth presupposes consecutive (though overlapping) historical phases that start out with invention and development, and move on to technology transfer, system growth and system stability, or maturity. These system-building phases do not apply to the case of Johannesburg. The early electrification in Johannesburg is better described by the expression *technological scramble*. *Technological scramble* uses the metaphor of a 'scramble for Africa' to emphasize the fierce and aggressive disposition of the electrification process outside the territories of the North Atlantic, which goes beyond mere industrial or economic competition and involves the contested realm of national powers and their spheres of sovereignty. The term is more fitting than technological growth, because the word 'scramble' designates 'an eager or uncontrolled and undignified struggle with others to obtain or achieve something' (Stevenson, 2010: 1598). The word conjures images of frantically climbing uphill on one's hands and knees. The technological scramble is best introduced together with the concept of the *settler company*. The technological scramble is performed by frontineers under the patronage of settler companies. These companies typically perform two strategic actions: *texploration* and *texpansion*.

Settler companies are controlled by frontineers. Unlike ordinary companies that function to make, buy or sell goods or to provide services in exchange for money, the goals of the settler company include the appropriation of certain rights and privileges that typically apply to places outside the social, political and jurisdictional order of the frontineer's home country. Because of its legal liability and obligations to distantly administered directives, the settler company carries with it a baggage of inherent sovereign capacities. These are not visible and often not accessible to the inhabitants of the region in which the settler company conducts its business. The settler company often deploys these sovereign capacities typically by means of technological enterprises.

The concept of the settler company is proposed here as a unit for historical analysis of technology. To examine technology through this lens helps us to move away from focusing predominantly on a legal and economic conception of companies to shed light on objectives and operations that remain unseen from such a point of view. For example, out of the many companies established in the name of Thomas Edison between 1878 and 1883, Hughes' book mentions those that pertain to his three case studies in Chicago, Berlin and London. Considered against his notion of technology transfer, the purpose of these companies was to transport and establish Edison's electric light and power station technology to other countries. In other words, the business of electrification required companies that were able to make, buy or sell technological goods or to provide technological services that relate to electric light or central power stations in exchange for money. However, the establishment of Edison companies displays significant patterns that are over-looked in this model of technology transfer. They claimed rights and privileges to colonial territories as well as contested territories not yet conquered by colonial powers. These companies imposed legal, political and administrative orders on these territories that paved the way for settlement by colonial powers.

In Hughes' image of networks, the pioneering role and power of 'companies' is not afforded adequate significance; instead, a company is regarded as an economic/financial means to spur the distribution of technology. The variety of companies involved in the electrification of Johannesburg, and their excessively broad spectrum of fields of competence, areas of jurisdiction and range of authorities, indicate that the notion of the company has served as a deceptive, administrative gap filler for a variety of practices and operations. In the case of Johannesburg, these practices by no means served the goal of electrification. Rather, electrification served the miscellaneous interests of these settler companies.

The fields of operations of the settler company are the frontiers of place and space – the material and epistemic conquest of territories of land and knowledge. For example, the company A. Goerz and Co. Ltd., which was financed by the Deutsche Bank, established Johannesburg's first central power station in 1897 and pursued the economic and industrial expansion of the German Empire in Southern Africa, a territory claimed by the imperial power of Great Britain. The German company A.E.G. assessed the possibility of harnessing electric power at the Victoria Falls at the beginning of the 20th century, before having any contract or potential client for electric power. The project was equally an opportunity for the A.E.G., and its employee Georg Klingenberg, to push the frontiers of large electric power systems. The British South Africa Company sought to use the electric power scheme on the Victoria Falls to further its imperial programme to conquer and settle the country (railways had only just arrived there) and to push the frontiers of the British Empire. This blending of settler companies' interests in exploration and expansion provided the electrification project with unwavering power – despite the speculative idea behind the project, which left open many technological questions. After the Victoria Falls electrification project was dropped, the original project was simply amended to serve a different goal for the settler companies involved – to establish a monopoly for the generation and supply of electricity to the gold mines of the Witwatersrand. The VFTPC challenged national, industrial, corporate, property, political, technical and legal frontiers. Settler companies involved in the electrification business selected their projects based on criteria that relate to texpansion and texploration, activities that both require the sense of technological entitlement of the frontineer.

Settler companies supply transitory shelter to the frontineer, the short-lived refuge to devise and direct his projects. Frontineers and settler companies do not follow system goals, identify reverse salients or solve critical problems – indeed, they adapt their goals and companies in a way that best positions them to derive territorial profits. Again, this adaptation is fuelled by a sense of entitlement, a sense of appropriation and a sense of intellectual ascendancy – the foundational drivers of electrification.

Exploration designates an expedition, an organized travelling act into unacquainted regions in order to gain material resources or knowledge. It includes activities such as sampling, mapping and prospecting, and it typically happens outside of the home country of the explorers. *Texploration* signifies an adaptation of this word, where the 'T' for 'technology' implies the impossibility of excluding technology from exploratory activities. A number of settler companies involved in the history of electrification in Johannesburg readily illustrate this kind of organized travelling project activity: the Edison Indian and Colonial Electric Company, Goerz, Siemens and Halske, the Rand Central Electric Works (RCEW), the General Electric Power Company (GEPC), the A.E.G., the BSAC, the VFTPC, the Rand Mines Power Supply Company and Merz & McLellan. The London Exploration Company Ltd. patently symbolizes this compulsive synergy. Texploration does not displace the idea of technological growth, but it gives it a direction, purpose and sense of conquest. However, it invalidates the ideas of the system goal, because it replaces the image of an ultimate goal that serves to identify and tie together the system components with a messy picture of a variety of technological projects of frontineers and settler companies competing for 'undiscovered' material and epistemic frontier territory. In addition, it disqualifies the simple idea of technology transfer as a neutral dissemination of technology across nations under mutually beneficial financial and material arrangements. The Briton George Forbes, consultant to the Power Scheme of the Niagara Falls, professionally considered the idea of harnessing the power of the Victoria Falls in 1894 at the request of Cecil Rhodes, and he travelled to Southern Africa for this purpose. This project was tied to Rhodes' plans to create a British Empire from the Cape to Cairo, which envisioned conquest of new territories by texploration through railways, 'white settlement', mining activity and electric power.

Texploration is intricately tied to *texpansion*, which emphasizes the act of territorial conquest over its prospect or vision. In Hughes' view, expansion of electric power systems '[…] was not simply an aggressive drive for undifferentiated size; it was a purposeful move to lower the cost of energy' (Hughes, 1983: 463). This economic argument, however, is not confirmed in the case of Johannesburg. Siemens and Halske's construction of the first central power station in Johannesburg was not motivated by the wish to lower the cost of energy. Siemens and Halske acquired a concession to produce and supply electricity on land in Brakpan to expand their business in Southern Africa and to gain territorial primacy in a country undergoing fierce imperial competition. Likewise, the idea to harness electric power from the Victoria Falls had less to do with the economics of power supply than with the idea of expanding A.E.G.'s business in Southern Africa and with conquering and settling the new British territory of Rhodesia.

Expansion denotes 'the act of becoming bigger or of making something bigger (territorial, economic), the process of increasing in size and filling more space, the process of making a business, organization, or activity grow by including more people, moving into new areas, selling more products etc., spread, when something increases in size, range, amount'.[214] Texpansion intends to emphasize that the movement of technology across borders right from the start requires a synergy with expansionary intentions. Again, projects of texpansion are quickly discernible in the activities of key players in the history of electrification in Johannesburg; Goerz, Siemens and Halske, A. E.G., Deutsche Bank, Dresdner Bank, Consolidated Gold Fields, BSAC, VFTPC, and so on. Texpansion also implies the requirement of the settler company.

214   http://www.macmillandictionary.com/dictionary/british/, accessed 3.2.2015.

The framework of frontineers, settler companies, technological entitlement, texploration and texpansion offers an alternative to the cultural forces that Hughes views as a formative influence in technological change.

> The cultural forces influencing the systems stemmed from the societies within which the systems grew. These societies were of various kinds according to time and place [...]. The cultural forces varied from society to society but there were also forces that transcended local or regional characteristics. These were mostly economic in nature. These economic forces in turn manifested the values that transcended time and place and pertained to Western society, or at least to the United States, Germany and England. The values were those of a cost-accounting, capitalistic civilization (Hughes, 1983: 462).

Hughes recognizes two sets of forces that influence electric power systems: the cultural forces of local and regional societies, and the economic forces of Western society, particularly the United States, Germany and England. What were the cultural forces that influenced electrification in Johannesburg between 1880 and 1930? The cultural forces influencing electrification in Johannesburg did not stem from the society within which the regional system grew. Johannesburg was a melting pot of migrants: migrant African labourers from all over Southern Africa, migrant Britons, migrant Germans, migrant Americans, migrant Dutch and migrant Boers. There was no single local society; at best, there was a myriad of local societies. Even then, the categories by which the composition of such local societies was determined reflected the views of the selected communities who claimed racial superiority. The gigantic electric power scheme of Johannesburg was built within just over two decades of the discovery of gold, in a newly explored gold mining region that required the forceful suppression, driving away or resettlement of local, indigenous populations.

Additional forces acknowledged in Hughes' model 'transcend' local, regional and national cultures and groups. He labels these forces 'Western society', but Hughes does not define this expression. Neither does he relate this expression to his systems framework or case study methodology. Clearly, Hughes assigns the United States, Germany and England strong positions in this category. The only indication that Hughes gives us about his conception of 'Western society' are the industrializing countries and regions that he mentions as forming the original pool of possible case studies: 'France, Italy,

Sweden, the Benelux countries, Russia, Japan, and other industrializing regions of the world' (Hughes, 1983: x).

In effect, by applying the expression of 'Western society' in this way, Hughes performs a particular act of writing history that has been described by various authors. Hughes' history of electrification in Western society establishes the myth that electrification began in the North Atlantic and from there spread to other societies where the technology was adapted into local and regional social, political and economic formations. The analysis presented here, however, has shown that this picture is faulty and that visions of global expansion and empire drove the project of electrification from the very beginning.

From the outset, Edison conceived his electric light and power station technology with a view to the global expansion of certain profitable rights and privileges. Indeed, because Edison cannot personally be credited with designing all of the parts that eventually came to compose these schemes (as Hughes also acknowledges), it might have been this strategy, which entailed establishing settler companies to participate in the technological scramble, that afforded him success. This technological scramble included international exhibitions and demonstration plants at sites that showcased the imperial and industrial power of the North Atlantic for its societies. The grand power scheme devised by Georg Klingenberg to provide the mostly foreign-owned gold mines of the Witwatersrand with cheap electric power appeared as one of two 'illustrative examples' in his textbook on large electric power systems. A gigantic electric power scheme that had been imposed with little resistance was classified as successful on economic and technical terms; indeed, it was an instructive model, with no regard to its broader impact on the political, social, economic and cultural lives of the people living in this region. Such a judgement was made from within a 'Western' frame of reference, with its attendant measures of attainment and triumph.

Viewed from this global perspective, Hughes' historical arrows of technological change – with their home position in the United States and spearheads shooting out to 'Western society', and from there on to the Rest of the world – lose their explanatory power for the history of electrification. Hughes' systems approach represents a particular view of geography 'in which the world has a single center, Europe [...] in reference to which all other regions are to be located; and an understanding of history in which there is only one unfolding of time, the history of the West, in reference to which all other histories must establish their significance and receive their meaning' (Mitchell, 2000:7). This 'modern' geography rests upon a particular order that is centred on Europe and is shaped by 'historical time, the time of the West' (Mitchell, 2000: 7). It 'presupposes an underlying unity in reference to which [...] variations can be discussed' (Mitchell, 2000: 24), and other histories 'tend to become variations on a master narrative that could be called "the history of Europe"' (Chakrabarty, 1992: 1).

Johannesburg is predestined to appear in Hughes' history of electrification as a variation of the master narrative of his systems model and its themes. The story of the transfer of Edison's technology to Berlin and London – which overlooks the simultaneous movement of Edison's and other technology – casts these specific histories of electrification in Europe as the primary subject of electrification, 'the historical residual against which difference is mapped' (Chakrabarty, 1992: 2). This kind of historiography contributes to the workings of Europe as 'a silent referent in historical knowledge' (Chakrabarty, 1992: 2).

> 'Only 'Europe', the argument would appear to be, is theoretically (at the level of the fundamental categories that shape historical thinking) knowable; all other histories are matters of empirical research that fleshes out a theoretical skeleton which is substantially 'Europe' (Chakrabarty, 1992: 3).

Chakrabarty speaks about a 'transition narrative' that 'situates the modern individual at the very end of history' (Chakrabarty, 1992: 10).

This modus operandi is remarkable in two respects. Hughes does not discuss the implications of his particular choice of case studies for this model of technological change – three great North Atlantic industrial powers of the late 19th and early 20th century. On the contrary, he claims that his model presents themes and subthemes that are valid for other cases as well. However, the United States, Germany and Great Britain occupied very specific positions in the family of nations between the years 1880 and 1930. They were confidently positioned at the frontiers of technological change. This unreflected specificity silently affords these countries a prime position as archetypes of technological change. Applying Hughes' 'method of growth analysis', which includes reverse salients and critical problems, outside of

these historically specific sites is predestined to generate histories of electrification that become variations and replicas of these archetypes. Thereby,
Hughes quietly places the North Atlantic at the centre stage of history, as an
'underlying unity' (Mitchell, 2000: 24) for the master narrative of technological growth. In this narrative, the North American system builder represents Chacrabarty's 'modern individual at the very end of history' (Chakrabarty, 1992: 10).

When this model is applied to the history of electrification in other
places, Hughes notion of 'Western society' comes to function as what
Chakrabarty (1992: 2) has termed the silent referent of historical knowledge.
Hughes' concepts of the reverse salient, the critical problem, the system
builder and technological momentum all contribute to establishing the silent
historical authority of this master narrative of electrification. When these
concepts are employed in other empirical sites, technology comes to be a
North Atlantic universal, always suggesting 'even if implicitly, a correct state
of affairs—what is good, what is just, what is desirable—not only what is, but
what should be' (Trouillot, 2002a: 221). In this way, Hughes' book sets the
terms of the debate on electrification at a global scale and restricts the range
of possible narratives on technological change.

Hughes evocation of the undefined category of the 'West' has definite
consequences for writing the history of electrification. These consequences
have to do with the way in which this category is constituted. The 'West'
acquires definitional meaning in relation to its residual opposite, the 'non-
West'. This residual opposite is embodied in the utopian image of 'the savage'. Trouillot describes this pair of utopian terms as the two faces of Janus.
These faces, however, appear in a chronological sequence that 'reflects a
deeper inequality: The utopian West is first in the construction of this complementarity. It is the first observed face of the figure, the initial projection
against which the savage becomes a reality. The savage makes sense only in
terms of utopia' (Trouillot, 1991: 30).

Hughes' history of electrification provides a textbook example of the
silent co-constitution of Western society and 'the savage'. By applying the
category of Western society, he categorically relegates the 'Rest' into the geographical opposite in place and space, and backwards in historical time. This
silencing or negation of the process of creating 'Western society' gives his
historical account definitional power over other historical approaches to

electrification. Because of this silencing, the category of 'the native' appears disconnected from the components that form part of the system or the network, as well as from its goals, style and classifications. The 'West's vision of order' requires 'two complementary spaces, the here and the elsewhere, which premised one another and were conceived as inseparable' (Trouillot, 1991: 32). Of particular significance to Hughes' history is the connection between the construction of an 'idealized image of civil society in the West' and its centre stage in history on the one hand, and a 'marginal' African society which is 'measured and found wanting' on the other hand (Mamdani, 1996, quoted in Randeria, 2002: 291).

The 'savage', which is part of the history of electrification in Johannesburg as a recurring theme in the form of 'the native', does not appear to relate to electrification, or technology at large, in any common-sense manner. On the contrary, however, these expressions appear to pertain to completely different worlds of relevance. The 'native' has nothing to do with Hughes' ingenious system builders of the industrial world, who solve critical problems that appear as reverse salients in order to achieve system goals – in and for 'Western society'. What has 'the native' to do with the history of electrification?

In the history of electrification of Johannesburg, 'the native' is a highly flexible category associated with gold mining, labour, exploration, civil society, governance, legislation and technology. This category served to justify the frontineer's projects of texploration and texpansion. It provided the necessary reasoning that legitimized the suppression and relocation of indigenous people, as required for gold mining, which in turn necessitated electrification. Walter Rathenau, a member of the board of directors of the A.E. G., explicitly formulated this necessary precondition for the colonization of the land of the native to legitimize 'corporate expansion and strategies of racial control' (Tuffnell, 2015: 54).

The question that Hughes asked in 1983 with respect to exploring the history of electrification – 'how do technological systems evolve?' – now needs to be rephrased to: 'what terms of the debate are put into place by the notion of technology?' The key challenge is to recast the terms of the debate beyond 'Western society' and to expand the range of possible histories to disclose its 'hidden faces' and categories, such as 'the native' in Johannesburg.

The concepts proposed in this chapter are intended to provide a starting point in the development of a vocabulary that guides the analytical eye to this hidden face. The *frontineer*, the *settler company*, *technological entitlement*, *technological scramble*, global *texploration* and *texpansion* may seem to be rather exotic categories when considered against Hughes' systems framework. However, perhaps they will be able to contribute to a revised history of the electrification of Berlin, Chicago and London in which new features come to light that reveal the global scope of the electrification project and expose the workings of its history of technology as a *North Atlantic universal*.

3.4 Seamlessness

What lessons may be learned for the field of STS from the study of electrification in Johannesburg? What have the frontineer, the settler company, technological entitlement, technological scramble, texploration and texpansion to offer the study of technological change more generally? To consider this question, we need to relate these notions to Hughes' model of technological change and its significance in the field of STS. Hughes' model proposes two units of analysis: the system builder and his settler company. These two units represent his point of entry for undertaking empirical research on technological change. Hughes also specifies the sphere of agency of the system builder and his organizations. He uses the metaphor of the seamless web of technology and society to designate their field of action. This metaphor at the same time charts the intellectual contribution of Hughes' model for the STS community. Hughes proposed the systems approach as a version of the so-called 'interactive model', which intended to overcome the classic dispute between external and internal approaches. The seamless web of technology and society provided the necessary terrain for Hughes' system builders to freely move across the realms of politics, economics, and society, since they all formed part of the seamless web.

The frontineer offers an alternative perspective to Hughes' system builder for empirically studying technological change. A comparison of the system builder and the frontineer therefore offers a critical perspective on Hughes' seamless web of technology and society. Because the seamless web links Hughes' system builder to other approaches in STS, these considerations also allow a discussion of the implications of the new concepts for the wider field of STS.

Hughes' Seamless Web of Technology and Society

Hughes uses the metaphor of the seamless web to specify the systems model that he introduces in *Networks of Power* in an article entitled *The Seamless Web: Technology, Science, Etcetera, Etcetera* (1986). In this article he also connects his systems model to the work of fellow scholars and discusses some of its broader implications for the study of technological change. Hughes describes his systems approach as a version of the 'interactive approach', which attempts to overcome the division between internalist (concerned with the content of technology) and externalist perspectives (concerned with the context of technology).[215] In his view, Barry Barnes' 'interactive hypothesis' to science should also be applied to technology. This would replace 'inside and outside, content and context' with 'the interactive', which is concerned with interconnection (Hughes, 1986: 285).[216] Hughes

---

215 'Within the past decade, as historians of science and technology began increasingly to seek an explanation for change, they referred more often to the context, especially the social context, of science and technology. These historians relegated science and technology to the inside of a context. This brought forth a more complex history that accorded more closely with our sense of messy reality. Rarely, however, did historians precisely define the categories that they placed in the context, nor did they always explain how the context and the content were related. They often had to resort to labelling the context as "social", or to stringing out a list of analytical categories, including the social, political, and economic. This, however, merely substituted one set of high-level abstractions for another and left too much room for misunderstanding the nature of the social, political, economic, etcetera' (Hughes, 1986: 282–3).

216 'There are problems with the contextual approach espoused by historians of science and technology, many of whom are reacting against the internalist mode. Flaws in contextualism began to appear when historians of technology rejected the notion that science is the context of technology, or that technology is simply applied science. They proposed an interactive relationship between technology and science. This, then, raised the question of whether the relationship between technology and other so-called contextual factors, such as the political and the social, should be redefined as interactive. The same question

advises caution with the use of 'hard, analytical categories – such as technology, science, politics, economics and the social [...] if their use leads to difficulty in comprehending interconnection' (Hughes, 1986: 285).

Hughes proposes networks and systems as alternatives to these hard analytical categories. Systems or networks are inhabited by two sets of interacting entities, which Hughes refers to as 'heterogeneous professionals – such as engineers, scientists, and managers – and heterogeneous organizations – such as manufacturing firms, utilities, and banks' (Hughes, 1986: 281–2). These heterogeneous engineers (or system builders) and organizations are Hughes' preferred units of study.

> Historians and sociologists who want to organize their research and writing in accord with an interaction model might, therefore, choose as their subject matter system builders – such as inventors, engineers, managers and scientists – or the organizations over which they presided, or of which they were an integral part (Hughes, 1986: 287).

The advantage of the network and systems perspectives, in Hughes' view, is that they 'eliminate many categories in favour of a *"seamless web"*' (Hughes, 1986: 281). The 'disciplines, persons, and organizations' that inhabit these systems or networks 'take on one another's functions as if they are part of a seamless web' (Hughes, 1986: 281–2). Accordingly, the seamless web is best grasped by following 'inventors, engineers, managers, and financiers who have taken a lead in creating and presiding over technological systems' (Hughes, 1986: 285). These system builders are 'no respecters of knowledge categories or professional boundaries' (Hughes, 1986: 285). Hughes first encountered the seamless web in the notebooks of Thomas Edison, who 'so thoroughly mixed matters commonly labelled "economic", "technical" and "scientific" that his thoughts composed a seamless web' (Hughes, 1986: 285). Hughes literally extrapolates this image of Edison's thoughts into his outside sphere of action and influence.

---

was asked about science and its context. A way out of the constraints of contextualism and into an interactive mode is now posed by the use of the "systems" or "network" approach' (Hughes, 1986: 281–2).

In other words, Hughes protagonists, the system builders (or heterogeneous engineers), inhabit the systems or networks, which in turn form a seamless web. The metaphor of the seamless web comes to represent the essence of Hughes' model. The system builders' seamless web maps their movements and spheres of action.

Several system builders who appear in Hughes history of electrification in Western society also figure prominently in the history of electrification in Johannesburg. Four of these individuals particularly inspired Hughes' notion of the system builder: the American Thomas Alva Edison, the Germans Emil Rathenau and Georg Klingenberg, and the Briton Charles Merz. All of these individuals also lastingly influenced the course of electrification in Johannesburg. However, this connection is not mentioned in Hughes' history of electrification in Western society; Johannesburg does not form part of the seamless web of technology and society of these system builders. Thus, the analytical category of the system builder fails to deliberate on the historical influence of Edison, Rathenau, Klingenberg and Merz on electrification in Johannesburg, and its implications for their own work and actions. The concepts of the frontineer, the settler company, the technological scramble, technological entitlement, texploration and texpansion, which developed from the empirical context of electrification in Johannesburg, offer analytical categories to reveal this connection. Viewing Hughes' system builders as frontineers provides an opportunity to reflect upon the performance and limits of the seamless web metaphor, the core of Hughes' model.

Hughes' seamless web metaphor has been widely referred to in science and technology studies literature and beyond. Having the status of a metaphor, it has naturally lent itself well to different interpretations. The variety of different meanings this metaphor has been used to stand for since Hughes' article is impossible to summarize or categorize. Nevertheless, I submit that its positive reception and wide use indicates that it has supplied a convenient label with which to fill a vacuous conceptual slot required for the emerging field of science and technology studies.

The Seamless Web Metaphor in STS

The metaphor of the seamless web stands at the beginning of a particular strand of scholarship that came to be known as the 'new sociology of technology'. This new sociology of technology was constituted by three approaches: 'the social construction of technology (SCOT), systems (later known as Large-Scale Technological Systems (LTS)), and actor-network theory (ANT)' (Bijker & Pinch, 2012: xiv).[217] The seamless web was put forward as one of the common traits of these approaches (Bijker & Pinch, 2012: xiv). It served as the organizing principle for an edited volume in 1987 that would come to figure as a founding text for the three directions in the new sociology of technology: *The Social Construction of Technological Systems: New Directions in the Sociology and History of Technology* (1987), edited by the three scholars Wiebe Bijker, Thomas Hughes and Trevor Pinch. This book and their authors' further work was instrumental in instituting the social constructivist approach to the study of technological change. The common denominator of the volume's articles is Hughes' notion of the seamless web:

> The second common trait to be found is the shared appreciation for Tom's "seamless web" metaphor. [...] all three approaches embraced the methodological principle of paying attention to how the borders between the social and the technical were drawn by actors, rather than assuming that these borders are pre-given and static. This also brings out the common element of a constructivist perspective. (Bijker et al., 2007: xvii).

The key message of the volume was that 'technological systems are both socially constructed and society shaping' (Bijker et al., 1989: 51). Various expressions have been introduced to designate this co-construction. Bijker uses the term 'sociotechnology' to designate that technological change is neither purely social nor purely technical (Bijker, 2010: 67).[218] According to

---

217   Although Bijker and Pinch qualify that '[t]hese three approaches did not completely represent the state of "new" sociology of technology in the mid-1980s' (Bijker et al., 2007: xv).

218   'The "stuff" of the fluorescent lamp's invention is economics and politics as much as electricity and fluorescence. Let us call this "stuff" sociotechnology. The relations that play

Bruun and Hukkinen, 'Pinch and Bijker modelled their approach on the sociology of scientific knowledge (SSK). SSK criticized the older institutional sociology of science – which focused on the social and institutional context of making science – for not opening the 'black box' of science, meaning that it had failed to make the contents of science an object of study for sociology (Bruun & Hukkinen, 2003: 100–1).

Bijker, Hughes and Pinch's 1987 volume was concerned with '[...] searching for a language and for concepts to express their new understanding of technological change' (Bijker et al., 1987:13). The authors presented 'three different approaches to deal with this *seamless web of technology and society*' (Bijker et al., 1987: 10, my italics). Bijker later described three interpretations of the seamless web of technology and society in subsequent research: first, it was used '[...] as a reminder that non-technical factors are important for understanding the development of technology'; second, it was used to indicate '[...] that it is never clear a priori and independent of context whether an issue should be treated as technical or social';[219] and third, it was used in connection with the symmetry principle to replace the allegedly purely social and the purely technical with 'sociotechnology' (Bijker, 2010: 67).

SCOT, ANT and LTS moved on to develop their distinct sets of concepts and models of technological change. Despite the variety in interpretation, the metaphor of the seamless web has offered a common reference point to develop new concepts for the new sociology of technology. Examples are John Law's 'heterogeneous engineering', Michel Callon's 'actants' and Bruno Latour's nature-culture and human-nonhuman divides. John Law's concept of 'heterogeneous engineering' exemplifies this emerging conceptual cohesion when Bijker describes it as having 'appealed to everyone as capturing aspects of the seamless web and constructivism' during the preparatory workshop to the volume (Bijker et al., 2007: xvii). Michel Callon introduces

---

a role in, for example, the development of the fluorescent lamp are thus neither purely social nor purely technical—they are sociotechnical' (Bijker, 2010: 67).

219 'The recognition that all kinds of social groups are relevant for the construction of technology (unit of analysis: artefact) and that the activities of engineers and designers are best described as heterogeneous system building (unit of analysis: technological system) supports this second usage of the seamless web metaphor' (Bijker, 2010: 67).

the term 'actants' as a higher level abstraction to designate the 'heterogeneous entities' that constitute a network (Hughes, 1986: 287). He criticized the categories or compartments of 'the elements in a system or network' and maintained that 'the fabric has no seams'. His concept suggested a higher level abstraction to address the problem of the hard categories of science, technology, economics, and politics, since 'these elements are permanently interacting, being associated, and being tested by the actors who innovate' (Hughes, 1986: 287).[220] Latour, in turn, has used the seamless web metaphor to designate the 'seamlesss fabric' of the 'nature-culture' and human-nonhuman divides for ANT. He applies the metaphor to present his holistic argument on how to analyse the 'modern' world: the modern fabric 'is no longer seamless' and 'analytic continuity has become impossible' (Latour, 1993:7).

The Seamless Web of Technology and Society in Johannesburg

The Edison Incandescent Lamp is unequalled for domestic and general illumination, and is suitable for all places where a steady, brilliant light, absolutely safe in its production and use, is required. It is peculiarly adapted for India and other tropical countries, as it will be unaffected by Punkahs and wind currents, and emits comparatively little heat. When once fixed it requires no skilled labour, but can be attended to by an ordinary native or other servant.
Prospectus on the Edison Indian and Colonial Electric Company,
Edison Papers 1882: (D-82-40x); TAEM 63: 32.

Hughes' notion of the seamless web can be studied by applying the perspectives of the system builder and the frontineer to individuals who influenced the course of electrification in Hughes' case studies on Chicago, Berlin, and London, and the case study on Johannesburg.

---

[220] Hughes refers to the article by M. Callon, 'The State and Technical Innovation: A Case Study of the Electric Vehicle in France', Research Policy, Vol. 9 (1980), pp. 358–76.

Thomas Edison

Thomas Edison was Hughes's favourite example to illustrate the seamless web (Bijker & Pinch, 2012). Hughes described Edison as 'moving between his Menlo Park laboratory and Wall Street to invent technology and raise capital in a seamless way' (Bijker & Pinch, 2012: xvii). Hughes' portrait of Thomas Edison as system builder accentuates his ability to assume a variety of roles at the same time, attending to issues of technology, economics, finance and management in parallel. Edison's proficiency in connecting different components, such as his 'technical' laboratory with the 'financial' district of New York, aimed to build a system that serves the system goal of technological growth.

However, if, we look more closely at Edison's 'seamless way' of moving about to invent technology, 'Wall Street' comes to signify more than mere financial matters. Wall Street was instrumental in shaping the legal and financial terms and conditions in the United States and the emerging global political economy of the New Imperialism of the 1880s. Edison did not simply connect his laboratory or technological business ideas with the Wall Street firm Drexel, Morgan & Co. to finance his electric light and power station project, or to secure his access to the British and continental market of patent rights. Instead, he sought a point of entry into the global architecture of the power of the day – admission to default privileges and exploitation of the British dominions. In other words, Edison exhibited a sense of global technological entitlement that motivated him to participate in the global technological scramble that trailed the imperial world order of the day. A simple geographic map of Edison's foreign companies charts this global sense of imperial entitlement to rights and privileges in far-away territories. Edison might have simply restricted his patents and companies to the United States, the prospective market of which could reasonably have been expected to yield large financial profits. Instead, Edison chose Great Britain, the most powerful Empire of the time, in an effort to ride that country's wave of industrial, financial and cultural success to establish his patent empire.

Edison's movements between his laboratory and Wall Street served this grand global scheme. Hughes judges this movement as illustrative of the seamless web of technology and society. Edison's grand global scheme is overlooked as a historical force in the 'interactive mode' of Hughes' model of technological change. It falls through the seamless web. Why? The Edison Indian and Colonial Electric Company will serve as an example to illustrate the reasons and to draw conclusions about the assumptions that underlie Hughes' metaphor of the seamless web.

The Edison Indian and Colonial Electric Company was formed in 1882 'for the purpose of acquiring and using in the Empire of India, Ceylon, Australia and South Africa, the rights and privileges of Mr. Thomas Alva Edison, relating to the application of Electricity of Magnetism as a lighting, heating or motive agent'.[221] The Company was established as a 'parent company' that would 'be prepared to undertake the business of a Lighting and Power Company, by lighting up towns, public buildings, manufactories, barracks, and residences, and supplying power by means of electricity; and in order to enlist local capital and influence will grant licences to Corporations and local Companies to carry on a similar business'.[222]

It is worth spending a moment to reflect upon the sites that are to be lit up by this Lighting and Power Company, because they may be assumed to form part of Edison's seamless web of technology and society. Whose towns, public buildings, manufactories, barracks and residences in sites of colonial rule were to be lit up by Edison's technology? The reference to 'the ordinary native' in the quote at the beginning of this section indicates that indigenous inhabitants were not seen to be beneficiaries of the technology. Instead, they were referred to in a functional role as labourers serving the colonial society: 'ordinary native' or 'servant', operator of the punkha.[223] The seamless web of technology and society only extended to a particular society, the British colonial society. In such colonial settings, Hughes' category of local society loses its meaning.

If we adhere to the systems approach of a seamless web of technology and society, we are forced to conclude that Edison's company disregarded the

---

221   'Except the application thereof for the purpose only of locomotion on railway or tramways or common roads, and except also the right of the Cape Government to use an installation which has been already sent out to them [...]' Edison Papers, 1882: (D-82–40x); TAEM 63: 32, pp. 1.

222   Edison Papers, 1882: (D-82–40x); TAEM 63: 32, pp. 2.

223   The word punkha was used in British colonies to name a fan attached to the ceiling and operated by a servant.

'ordinary native or other servant', as a member of society in the British sovereign territory of British India. The 'ordinary native or other servant' is only referred to in his labour-performing capacity. This is by no means a historical detail; it is emblematic of Edison's vision of the seamless web. The web only allowed certain peoples to move about in a seamless way. It simultaneously banned other peoples' movements outside of this web. This mechanism relegated people categorized as 'the ordinary native' to historical invisibility. Technology was the key to allow this exclusive free movement for some and to deny it to others. The Edison Incandescent Lamp was designed to illuminate domestic sites of specific societies, including the ruling British society in the colonies but excluding local inhabitants. In this way, Edison's project contributed to establishing the norms for the 'society' for which his technology would provide solutions to critical problems. Technology does not only function in harmonious interplay to constitute society; it also cements a restricted normative image of 'Western' society that produces images of difference, of populations aspiring to move up the ladder of civilization to achieve the status of society.

Edison, the hedgehog, knew one big thing, connected by a seamless web, but the hedgehog moves within a confined habitat. His short legs do not allow him to walk beyond the boundaries of his home territory. If he imagines the world beyond this habitat, it is likely to be an extension of the colours and shades of his seamless domestic web. Edison filed patent letters for his ideas and technologies in far-away territories that he had never visited; they were fictions of his imagination. Even in legal terms, the contours of these territories remained vague. Edison's attorneys in 1889 were unable to establish whether the 'South African Republic' fell within the confines of their rights and privileges. These territories existed in Edison's geography of imagination, they inspired his technological entitlement that induced his technological scramble to establish and manage his technological rights and privileges in these territories. The fragile status of territories at the outskirts of the North Atlantic geography of imagination permitted a more unproblematic and direct application of the geography of management to impose the seamless web of technology and society. In other words, Edison's particular geography of imagination engendered the geography of management that he invented to expand his imagined seamless web of technology and society.

To expand the seamless web, Edison presented his technology at the nodes of power of a world of empire: Wall Street in New York, the Paris Electrical Exhibition at the Champs-Élysées in the Palais de l'Industrie, and the Holborn Viaduct in London. His particular choice of sites to exhibit his technological prototypes served the texploration and texpansion of the frontineer Edison and his settler companies rather than a straightforward and neutral notion of 'technology transfer'.

However, Edison's technology was not simply inserted cleverly into an existing imperial world order – the period of New Imperialism had only just begun. Edison played a part in shaping this period – his patent and company expansion across the globe contributed to the emerging international patent and investment law. His imagined seamless web of technology and society fitted well into the aspirations of these projects.

Emil Rathenau

Hughes presents Emil Rathenau as another system builder in the history of electrification. Rathenau's name 'looms large in the history of the electrical industry in Germany and also in the establishment of Berlin and its Elektropolis' (Hughes, 1983: 179). Hughes refers to Rathenau in his capacity as 'head of Allgemeine Elektricitäts-Gesellschaft, Germany's largest manufacturing, electrical utility, and banking combination before World War I' and notes that 'he envisaged the entire German economy as functioning like a single machine' (Hughes, 1986: 286). In Hughes' view, 'Rathenau stood for the powerful and widely influential interaction of investment capital, industrial enterprise, and highly organized marketing' (Hughes, 1983: 179).

In Hughes' narrative, Rathenau is portrayed as presiding over the process of transferring Edison's technology to Berlin and Germany. He fitted the technology to suit local cultural, political and economic conditions so as to fabricate a seamless web of technology and German society. For this purpose, he acquired the Edison patent rights for Germany to establish the Deutsche Edison-Gesellschaft für Angewandte Elektricität in 1883, and the A.E.G. in 1887. In this portrait, Rathenau's sphere of operation is Germany. The seamless web that he constructs envelops the territory of the German Empire, its industries, politics and society. According to Hughes, Rathenau travelled to America in 1903 to '[reach] an agreement with General Electric in America to divide their world markets'. The result of this agreement was that the 'A.E. G. would continue to be preeminent in Europe: [General Electric], in North America' (Hughes, 1983: 179).

Viewing Emil Rathenau as a frontineer sheds a different light on Rathenau's trip to America. From this perspective, Rathenau travelled to America to represent the interests of the A.E.G., a settler company intent on pushing its territorial frontiers. These territorial frontiers reached way beyond 'Europe' and 'North America'. Rathenau's global imperial aspirations are formulated in a letter to Edison (who had written to request that the A.E.G. not compete in Japan) dated 19 February 1889: 'You have not taken into consideration the fact that we have during the last year entirely discharged all the duties toward your rightful successors in Europe [...] [A]nd since we have accomplished this *we do not stand any longer*, as you describe it, as one of the Edison companies, *confined to a certain territory* [...]. The market for our product is *the entire world*' (Hausman et al., 2008: 79; my emphasis).

The fact that Hughes only mentions the categories of Europe and North America in this process of dividing the world market between the A.E.G. and General Electric betrays his own geography of imagination: his seamless web only recognizes, and therefore connects, technology and society in the North Atlantic. In actual fact, Rathenau's system-building activities reached far beyond this geographical area.

Emil Rathenau was one of the key initiators of the VFTPC's grand power scheme in Johannesburg, the grand power scheme constructed by the A.E. G. and owned by German banks and the British South Africa Company. Rathenau served as one of the VFTPC's two German directors until the First World War. The great significance of the power scheme on the Rand for the A.E.G. is evidenced by the fact that Rathenau reported to the German Emperor in 1906 on the imminent contractual agreement between the A.E.G. and the British South African Company. He portrayed the the VFTPC contract with the A.E.G. as British acknowledgment of German superiority (Weinberger, 1971). This portrait testifies to the sense of technological entitlement that fuelled his involvement in the electrification of Johannesburg. This involvement was not motivated by a system goal or reverse salient; it represented texploration and texpansion that served the technological scramble of global electrification. Incidentally, the A.E.G.'s business operations in

Johannesburg brought the settler company substantial profits that buoyed their finances in the critical years of the first decade of the 20th century.

Georg Klingenberg

Hughes' model system builder in Germany is Georg Klingenberg, 'the engineering head of the A.E.G.'s power plant design and construction division, an engineer of international reputation and author of definitive works on power-plant design and operation' (Hughes, 1984: 197). In Hughes' framework of system builders, Klingenberg appears as one of the key figures in the history of electrification in Western society. Hughes notes his acquaintance with the British system builder Charles Merz and mentions that Klingenberg also testified to the British Parliament on the eve of the First World War regarding London's backward state of electrification and the need for large central power stations (Hughes, 1983: 228).

Klingenberg travelled to Johannesburg in 1906 to conduct negotiations that would lead to the establishment of the VFTPC and his design of 'one of the largest power works in the world'. He later gave evidence before the Power Companies Commission of the Transvaal Republic in 1909, which had been convened to look into the need for government regulation in the face of the VFTPC's rising monopoly and plans to establish a grand electrification scheme to supply electricity to the gold mines of the Witwatersrand. Klingenberg testified here in his capacity as an internationally renowned engineer, but he was also the head of the A.E.G.'s power plant design and construction division – the company that would receive the contracts for the construction and equipment of the grand power scheme. Four years later, in 1913, Hughes published a textbook on *Large Electric Power Stations: Their Design and Construction, with Examples from Existing Stations*. This book dedicates more than 100 pages to introducing the model example of the power system of the VFTPC (the other example is the Märkische Electricity Works). At this stage, the A.E.G. had not yet managed to realize the idea of a regional power system for Berlin. In the same year, Klingenberg presented a paper to the Institution of Electrical Engineers in London on 'Electricity Supply in Large Cities'. The paper, which compares the electricity supply in London with Chicago and Berlin, is extensively referenced in Hughes' book.

To realize this power scheme some 5,500 miles south of Berlin, Klingenberg undertook a series of negotiations with various stakeholders, such as the
A.E.G., the British South Africa Company, the Deutsche Bank and other
German and Swiss banks, the gold mining companies of the Rand Mines/
Eckstein Group, the Transvaal government, the municipality of Johannesburg and the Coal Owners Association. Assuming that these negotiations
chart Klingenberg's movements through the seamless web of technology and
society, his path highlights the specific features that led to successfully managing the installation of Johannesburg's grand power scheme. Although the
particular technology of Klingenberg's seamless web is widely documented,
we know little about its society. What kind of society is represented by the
institutions that were involved in the negotiations? Whose interests do these
institutions represent?

Most of these institutions (certainly the ones that eventually would
stand to gain from the project) represented foreign financial and imperial
interests. There is no sensible way of claiming that these institutions represented the interests of any kind of society other than the ones involved in the
imperial scramble for the Transvaal territory. In addition to the British and
German societies, this included the Boer society and the society of North
American expatriates (such as American mining engineers) in Johannesburg,
all belonging to a racial category defined by skin colour. It might be argued
that these institutions only represented particular segments of these societies,
but the point to be made here is that they certainly excluded the majority of
the inhabitants of the Transvaal at the time. The 'natives' did not belong to
any kind of society. Therefore, they were considered neither in Klingenberg's
negotiations nor in the seamless web that he imagined for his grand power
scheme.

However, there was no seamless web of technology and society in
Johannesburg at the time. The population of Johannesburg consisted of
migrant communities from Southern Africa, Europe and other places. Klingenberg's negotiations were not only about the economics and finances of
technology; they were also about constituting a particular kind of society: his
ideal image of the seamless web of technology and society realized in a grand
technological system that served the establishment of white settler society in
Johannesburg. Accordingly, his negotiations concerned issues related to the

purchase of land, concessions, industrial monopoly, imperial politics and native labour.

The VFTPC was co-financed by German banks and the British South Africa Company, a chartered company under the British crown with extensive rights to conquer and exploit territories in Southern Africa. This joining of forces is not simply symbolic of the mutual constitution of technology and society; it is emblematic of how the myth of the seamless web was used as a strategic tool for designing this mutual constitution to further particular political and economic interests. To constitute the category of society, however, it was necessary to remove from the web those populations that disturbed the seamless picture and revealed ruptures, ambiguities and inconsistencies, such as those embodied in the utopian idea of 'the native'.

However, even if the connection between technology and the 'native' populations is conceptually cast out from the seamless web of technology and society, it is present in the historical record. In fact, it appears in the considerations of the system builders. For example, Klingenberg's account of the model case of the VFTPC spells out the direct relationship between electrification and labour.[224] Klingenberg also comments on the 'exceedingly simple' legal conditions for the property rights and concessions necessary to attain official permissions for the overhead and underground cables for the transmission of energy on the Rand. (Klingenberg, 1916 [1913]: 239). The legal conditions as they appeared to Klingenberg only appeared as 'exceedingly simple', or seamless, because they disregarded the property and land rights of the majority of the population that inhabited the Transvaal.

The seamless web of Hughes' system builder Georg Klingenberg appears to be a strategy for effecting coherence, to institute smooth complementarity between technology and society. Viewed through the prism of the frontineer and the settler company, this strategy is inverted to reveal an underlying assumption of the systems perspective: the seamlessness of technology and

---

224   'At the time of the author's visit a law had come into force prohibiting the importation of Chinese labour, and decreeing the gradual substitution of the latter by local labour; further measures had been proposed for the prevention of double shifts on underground work. The influence which such a radical change in the system of working would have upon the output and load factor had, therefore, to be considered from the outset' (Klingenberg, 1916 [1913]: 201).

society. This notion is intended to capture these systematic erasures in the dominant historical narrative on electrification that are generated by the metaphor of the seamless web of technology and society.

Charles Merz

The British consulting engineer Charles Merz also appears in Hughes narrative as 'a system builder, a Hedgehog like Thomas Edison [...]' (Hughes, 1983: 249), whose entrepreneurial talents 'ranged from engineering to management and to politics and finance as well' (Hughes, 1983: 205). In Hughes' story, Merz's system building aspirations were forced to endure a long struggle in London. Hughes describes a series of ignored efforts on the part of Merz to build a unified electrical system in London.[225] Hughes attributes the failure to realize Merz's vision to the 'resistance of proponents of local government and the foes of private ownership' (Hughes, 1983: 205). In other words, because of their vested interests and lack of hedgehog perspective, the established institutions in London stood in the way of Merz's (rational and economical) efforts to unify the electric power supply and to realize the seamless web of technology and society.

In far-away Johannesburg, in contrast, the advice of the British consulting engineer Charles Merz was sought and heard. The Smuts Government of the newly established Union of South Africa was keen to solicit Merz's advice and appointed him to report on national electric power supply in 1919. Indeed Merz's recommendations were so well received that they led directly to the Electricity Act of 1922, which established Escom, the electricity utility that still occupies a monopoly position in today's South Africa. The report submitted to the South African government on behalf of Merz clearly presents his image of the seamless web of technology and society.

---

225 According to Hughes, '[...] the politicians were not receptive to Merz's plans' in London before World War I (Hughes, 1983:205) and Merz's proposed Bill to Parliament to establish a power company to develop a central power scheme in London was rejected in 1907 (Hughes, 1983:255). Merz's invited report in 1914 to the London County Council recommending to establish a company composed by the existing utility companies to coordinate the London supply situation was not taken further either (Hughes, 1983: 256).

The importance of developing the supply of electric power in the Union arises from the need for encouraging its existing manufactures and attracting new and permanent industries as rapidly as possible. This latter has been emphasized by yourself, and by many important public bodies, who have stated that *in no other way can an adequate white population be obtained*. It is therefore a policy which justifies the expenditure of money in advance of to-day's actual needs (Merz & McLellan, 1920: iii; my emphasis).

In Merz's vision, the seamless web of electrification technology and society in the Union of South Africa would serve the settlement of white society in the 'new country'. Accordingly, his recommendations for electricity supply to the Union of South Africa were geared to serve the requirements of the white settler communities. By default, electrification was understood to provide for the interests of select populations. Other populations did not qualify as societies; instead, by implication, they were viewed as inferior cultures and therefore did not form part of Merz's vision of a seamless web of technology and society that would develop the Union of South Africa.

In this example, the vision of the seamless web immediately and powerfully eradicates histories other than that of the white population from the stage of the world history of electrification. The new country of the Union of South Africa only comes into existence with the settlement of 'white' society. In this way, electrification and its technology come to constitute society – not society as an expression of specific South African political, economic and cultural features, but as a normative category that quietly installs racial standards of exclusion. In this manner, the seamless web of technology and society comes to exclude entire populations, or complex population formations, from historical writing. In 1905, the British Colony of the Transvaal had a population of some 300,225 'European', 1,030,029 'Aboriginal' and 23,946 'Coloured' people (Native Affairs Commission, 1905: 27 (Annexures)). To erase successfully from 'society' and history a population that constituted over three-quarters of the inhabitants has to be considered an extraordinary achievement by any measures. Accordingly, it is a difficult challenge to reinsert these populations into the dominant historical narrative because we have to sidestep established analytical and descriptive categories of silent exclusion while at the same time abiding by the academic standards that are set in STS.

Hughes is well aware of Merz's international influence and refers to him as an 'agent of technology transfer on a grand scale': 'Charles Merz travelled throughout the world for Merz & McLellan [...]. Merz also observed engineering practices, prepared reports, and organized construction projects in Australia, Argentina, South Africa, India and in American and British cities other than Chicago and Newcastle upon Tyne' (Hughes, 1983: 452–3). Hughes compliments Merz with this distinction because he '[took] with him the experience he gained in building the NESCO system on the northeast coast of England and [brought] back state-of-the-art ideas from the rest of the world' and because he 'associated with other leaders in electrical engineering and management, such as Georg Klingenberg of A.E.G.' (Hughes, 1983: 453). In Hughes' view, he 'articulated and disseminated the economic and technological principles of regional systems' (Hughes, 1983: 453). He concludes that 'no history of technology transfer would be complete without consideration of consulting engineers like Merz' (Hughes, 1983: 453).

Merz, in turn, considered the electric power scheme on the Witwatersrand as 'one of the most important in the world'. Indeed, he referred to the example of Johannesburg, where 'many of the largest mines obtain their power from a central supply rather than put down stations for themselves', as illustrative of the 'proof of the soundness of the policy of concentration' (Merz & McLellan, 1920: 8). On what basis did Merz pronounce this judgement? By what criteria does Merz appraise the success of the policy of concentrating large central power stations in Johannesburg?

Merz judges the case of electrification in Johannesburg in comparison to London. He contrasts these cases by the order of precedence of electrical legislation and industrial development that differs in 'older' and 'newer' countries. In 'Great Britain and the older countries generally', where industrial development preceded electrical legislation, the 'development of manufacturing' was impeded. In contrast, he considers South Africa to be fortunate to be able to 'lay down its power supply system almost at the foundation, instead of having to superimpose it upon an old-established industrial fabric' (Merz & McLellan, 1920: 19). Accordingly, his report recommended that, as a 'new' country, South Africa should adopt a policy and establish a system of power supply 'before the industrial development of the country has proceeded any further'. In other words, Merz recognized that electrification had free reign to compose the seamless web of technology and society in South Africa –

before institutions had grown to represent the various social groups and their interests.

From this perspective, Hughes' verdict on the backwardness of electric power supply in London at the beginning of the 20th century (which reproduces that of Merz, Klingenberg and Insull) appears in a different light. Hughes identifies 'vested and historical interests' as the causes of London's backwardness and suggests that they needed to be overcome by the 'progressive combination of coordinated forces'. He is persuaded that the reasons for this backwardness are political, or legislative, and he compares the situation with Berlin. However, these vested interests (Hughes, 1983: 260) might also be read as democratic forces of resistance to the system builder's projects that tried to impose the seamless web of technology and society without avail. The many institutions already established to represent both electrical technology and civil society revolted against the seamless web that did not fit with the seams already sown together.[226]

Hughes' notion of technological progress measures successful electrification in relation to the achievement of the system goal. He regards the social forces of resistance to technological change in London as an expression of the local culture in response to the rational proposals of the system builders, such as Merz. They impeded the building of 'bridges between the world of technology and that of politics', the combining of coherent, co-ordinated forces of technology with 'finance, industry, and utility' to create the system network. In other words, Merz and others did not succeed in their pursuit of large central power stations because of the seamless web of technology and society. In Johannesburg, these social forces of resistance and their institutions did not exist.

---

226   'In Berlin, there were individuals who built bridges between the world of technology and that of politics. The presence of bankers and industrialists on the advisory board of BEW was a manifestation of the coherence and co-ordination of finance, industry, and utility in Berlin. Such a progressive combination of coordinated forces was needed to overcome vested and historical interests in London, but such a combination did not exist. Instead, the proponents of local government authority, of municipal socialism, and of private enterprise confronted one another in a pluralistic debate that from the point of view of the forces for technological change produced a stalemate' (Hughes, 1983: 260).

Merz's report on the electrification of South Africa in 1920 is strikingly illustrative of the system goal of his envisioned large electric power stations – and his corresponding seamless web of technology and society. In this report, he refers to the segment of the Southern African population not included in the 'white population' but only in their capacity as 'valuable supplies of good native labour' (Merz & McLellan, 1920: 1).[227] However, the 'white population' category is neither clear-cut nor defined by any kind of natural circumstances or colour yardsticks. The co-constitution of electrification and (white) society served the technological scramble in which Merz's consulting company won several large consulting contracts, such as for the construction of the first central power station after the First World War in Johannesburg, the Witbank power station.

Hughes refers to Merz as an example to argue for local and regional differences in technological style. Referring to the overseas experiences of Merz, Hughes concludes that regional systems were shaped by cultural forces that 'stemmed from the societies within which the systems grew' and that 'these societies were of various kinds according to time and place [...]' (Hughes, 1983: 462).[228] Huges concedes that '[t]he cultural forces varied from society to society but there were also forces that transcended local or regional characteristics' (Hughes, 1983: 462). The large electric power

---

227 Merz and Klingenberg testified not only before the British parliament but also before the Commission of Inquiry into the Power Companies of South African Union in 1910 which led to the Transvaal Power Act by the Transvaal Colonial Government. Merz and McLellan's report begins with the sentences: 'In the development of every new country the first necessity is the establishment of means of communication – roads, railways, telegraphs. At a later stage, depending upon the labour, raw materials and markets available, manufactures spring up. In order to develop and flourish, manufactures nowadays need cheap power' (Merz & McLellan, 1920: 1). They consider South Africa to possess 'most of the requirements needed for the development of manufactures' and reckon that 'although at present its white population is not proportionate to its area and possibilities, it is probably sufficient for its immediate manufacturing needs if power be more widely used, while it obtains valuable supplies of good native labour' (Merz & McLellan, 1920: 1).

228 'In the 1904 article and later essays Merz stressed the regional character of technology. Because of his familiarity with technology throughout the world, he realized that a universal best way did not exist, instead, a variety of styles prevailed. He believed that local conditions completely governed power supply within a region' (Hughes, 1983: 454).

scheme of Johannesburg, however, did not stem from the societies within which the system grew. Instead it imposed a seamless web of society and technology that overlooked the majority of the population in favour of a particular conception of society. This conception followed racial criteria, such as skin colour, and therefore only considered a small proportion of the inhabitants of Johannesburg.

The seamless web as strategy

One of the common qualities of Hughes' system builders Edison, Rathenau, Klingenberg, and Merz is their shared vision of the seamless web of technology and society. In Hughes's study on 'Western society', the system builders of electric light and central power stations move about freely in the seamless web of technology and society. Here, electrification results from the interactions on this seamless fabric: between local societies, their cultural, political and economic institutions, and the system builders. Edison, Rathenau, Klingenberg, and Merz's involvement in the history of electrification in Johannesburg expands their seamless web beyond the boundaries of Hughes' 'Western society'. Their activities outside of the seamless web in these geographical regions, however, disclose that Hughes' metaphor of the seamless web carries a particular imaginary geography. The imaginary geographies of Edison, Rathenau, Klingenberg, and Merz inform their vision of successful electrification in Johannesburg. This vision prefers large central power stations. It views society as a normative category that contrasts 'civilized' societies with uncivilized cultures. As a result, electrification in Johannesburg may be seen as resulting from the extension of the frontiers of the seamless web of technology and ('civilized') society rather than from interaction with local society. The seamless web now appears as a metaphor with a specific vision and underlying assumptions.

The personalities that shaped the course of electrification in Johannesburg during the time span that Hughes regards as formative for power systems imposed this vision. In their enterprises, they were able to depict this relationship as mutually beneficial and unproblematic. Of particular note is their audience: they speak to European and North American societies. Other societies are projected backwards into historical time on the scale of civilization. The seamless web of technology and society does not span this historical time; it is designed to serve the frontiers of societies, the model society: European civilized societies. To impose the seamless web of technology and society on other, less civilized societies or cultures becomes a programme that does not simply transfer technology.

The seamless web identified by Hughes simply reflects the system builders' visions. His history maps their quest to institute a smooth path between technology and society, tailored, so to speak, by the seamless web.

The concept of the system builder assumes the metaphor of the seamless web. Indeed, one might read Hughes' theory of technological system growth as a story of how the seamless web expands. On the surface, this metaphor serves to chart the great sphere of movement and action of the system builder, and to soften hard analytical boundaries between technology and society. Below this surface, however, the metaphor of the seamless web also serves to naturalize contradictory actions and roles of the system builders, because it allows them to move about smoothly and to 'eliminate categories' (Hughes, 1983: 281). The seamless web provides an acceptable rationale for their deviation from the public roles, tasks and responsibilities with which these individuals are typically associated. In other words, the seamless web dissolves any kinds of questions that relate to paradoxical, unauthorized or otherwise puzzling conduct or attitudes that might arise from their involvement in 'physical artefacts, mines, manufacturing firms, utility companies, academic research and development laboratories, and investment banks' (Hughes, 1983: 287). The system builder moves about the imaginary seamless web of technology and society to effect technological growth and to pursue his imaginary system goals.

Viewed through the lens of the frontineer, however, the system builder assumes the seamless web as a strategy to justify certain actions and to conceal others. These actions are directed at pushing material and epistemic frontiers. Once these boundaries are expanded, the frontineer needs to fasten his new material and epistemic territories to his old assets. In other words, his core business is to sew seams together, to make things fit and to glue the new beyond the frontier to the old. Technology becomes the stitching for these seams. This forceful act is validated by imposing the imaginary assumption of seamlessness between technology and society. Rather than dissolving the boundaries between technology and society, the metaphor of the seamless web overplays the violent actions that are needed to pursue this vision. The ostensibly natural fit of a smooth passage between technology and society appears to be a strategy for constituting society in a complementary, harmonizing relationship with technology. The vision of the seamless web attests to this dual constitution. Hughes envelops the activities of system builders with a cloth of seamlessness that allows them to rush about unobtrusively and temporarily to take on roles and carry out activities with which they are not associated in the public sphere.

The Grand Assumption: Seamlessness

Hughes views the 'great technological systems, utility networks, trusts, cartels, holding companies' of the 20th century as seamless webs created by system builders (Hughes, 1986: 286). System builders may be identified by their ability to powerfully interconnect the diverse components of technological systems, such as 'physical artefacts, mines, manufacturing firms, utility companies, academic research and development laboratories, and investment banks' (Hughes, 1986: 287). System builders pursue a system goal. System components are defined by their participation in the pursuit of this goal: 'These components make up a system because they fall under a central control and interact functionally to fulfill a system goal, or to contribute to a system output' (Hughes, 1986: 287).

In Hughes history, the ultimate system goal of the system builders in 'Western society' was to weave electrification into the seamless web of technology and society. Technology transfer, the reverse salient, technological momentum and technological growth provided the necessary steps to reach the system goals. These and other concepts that Hughes uses to describe technological change (technological style and culture, system stability and maturity) configure the seamless web of technology and society for historical analysis. Based on an examination of Johannesburg, the above analysis has added new concepts to describe critical properties of the history of electrification that are eclipsed in Hughes' framework. These concepts challenge the assumption of a seamless web of technology and society. To capture this challenge and to render it effective for further analysis, the notion of '*Seamlessness*' is coined.

The term 'Seamlessness' is suggested here as an analytical tool for designating a set of unstated assumptions that have influenced the course and development of the field of STS since the mid-1980s. Seamlessness refers to the metaphor of the seamless web of technology and society as proposed by Thomas Hughes. This metaphor assumes that science and technology are surrounded by hard categories such as politics, economics, society and culture. These categories can be favourably transcended by applying the metaphor of the seamless web of technology and society. This elimination of categories provides a tabula rasa for analysis that naturally encloses all possible histories. In effect, however, the seamless web implies limited conceptions of 'technology' and 'society' and also constitutes their opposites.

This erasure of categories leaves behind the idealized historical residue of 'the West', or 'Europe', which maintains a unique and universal status. Technology comes to perform as a *North Atlantic universal*. Seamlessness establishes Europe as 'silent referent in historical knowledge' and casts a restricted theoretical condition for knowledge production (Chakrabarty, 1992: 2) in the field of science and technology studies.

Seamlessness can be exposed empirically by employing the units of study of the frontineer and the settler company. These units displace Hughes' image of the system or network as the arena for historical action in favour of a geography of frontiers. These frontiers map material and epistemic territories that can be studied through Trouillot's geographies of management and imagination. Among the projects of the frontineer is the technological scramble, which is spurred by his sense of technological entitlement. He typically engages in the activities of texploration and texpansion in the geographies of place and epistemic place.

The unstated assumptions of the seamless web carry out a strategic function in STS: rather than simply eliminating categories, they endorse categories of thinking that presume a seamless unfolding of historical time centred in Europe, the West. The seamless, harmonious co-constitution of society and technology is a means to the end of advocating the West and naturalizing the contradictory and at times forceful historical actions to pursue technological growth. Seamlessness designates a family of assumptions that have carved out territorial standards for STS that have persevered to the present day. For this reason, these assumptions have restricted the applicability of STS and have posed obstacles to the development of STS.

In the early history of electrification in Johannesburg, society is not an appropriate category for designating the diversity of populations affected by technological change. Neither does the category of culture help us to overcome this problem. No seamless web can be ascertained between technology and society. Instead, the category 'electrification' is escorted by its shadow category 'the native', to designate non-white, residual populations that are not captured by the word society. The 'native', however, remains in the background and shows no direct connection to electrification. He is defined by the colonial settler as a 'labourer' rather than as a member of civil society. Thus, he does not form part of the seamless web of technology and society. The category of 'the native' falls through the cracks of the seamless web and as a consequence vanishes from historical analysis. We lack the necessary concepts to resituate 'the native' in the history of technology.

The 'native' defies conceptual connection to the early history of electrification in Johannesburg, although the population groups subsumed under this category composed three-quarters of the peoples living in the Transvaal in the early 20th century. The Seamlessness concept is intended to draw attention to this erasing property of the seamless web of society and technology that banishes 'the native' from history.

The seamless web of society and technology engenders Trouillot's 'Savage Slot'. Trouillot claims that the idea of the 'West' 'structures a set of relations that necessitate both utopia and the savage' (Trouillot, 1991: 24). He views the 'West' as a utopian, universalistic project that is inconceivable without its complementary 'other'. In this projection, 'the savage becomes a reality', but 'the savage makes sense only in terms of utopia' (Trouillot, 1991: 30). Seamlessness labels the utopia of the seamless web: an ideal condition of a smooth, orderly alliance of technology and society. This project is inconceivable without its complementary 'other'. The native cannot form part of the seamless web because he is needed as its residual category. The native's function is to constitute the utopian society aspired to by the metaphor.

The literal meaning of the word *seamless* does not simply denote having no seams. It also designates an active property of moving from one thing to another easily and without any interruptions or problems, with no awkward transitions, interruptions, or indications of disparity;[229] or changing or continuing very smoothly and without stopping;[230] happening without any sudden changes, interruption, or difficulty;[231] without a break; smoothly.[232] In addition, it indicates a condition of perfection, having no flaws or errors.[233]

The orderliness aspired to by the dictionary definitions of seamless relates directly to the categorical omission of 'the native' as an inhabitant of the seamless web. Trouillot's Savage Slot describes the mechanism underlying this relationship:[234] the savage, or 'the native', is a prerequisite for the absolute order and utopia that lies beneath the idea of 'the West' and its imaginary geography of civilization.[235] Importantly, 'the West's vision of order implied from its inception two complementary spaces, the here and the elsewhere, which premised one another and were conceived as inseparable' (Trouillot, 1991: 32). The first regional power scheme of Johannesburg installed technology to constitute (white) society – and its opposite, or permanent aspirant, 'the native'.

Hughes' metaphor of the seamless web has been used in other strands of STS to erase various kinds of categories, but the ideal condition envisioned in the original metaphor has been retained as a silent but powerful assumption. Hughes' concepts for describing technological change do not simply con-

---

229 Merriam Webster, http://www.merriam-webster.com/dictionary/seamless.

230 Macmillan, http://www.macmillandictionary.com/dictionary/british/seamless.

231 Cambridge, http://dictionary.cambridge.org/de/worterbuch/englisch/seamless.

232 Collins, http://www.collinsdictionary.com/dictionary/english/seamless.

233 Merriam Webster, http://www.merriam-webster.com/dictionary/seamless.

234 'In the context of Europe, the works that set up these slots were part of an emerging debate that tied order to the quest for universal truths, a quest that gave savagery and utopia their relevance. Looming above the issue of the ideal state of affairs, and tying it to that of the state of nature, was the issue of order as both a goal and a means, and of its relation with reason and justice' (Trouillot, 1991: 31).

235 'Colonization became a mission, and the savage became absence and negation. The symbolic process through which the West created itself thus involved the universal legitimacy of power – and order became, in that process, the answer to the question of legitimacy. To put it otherwise, the West is inconceivable without a metanarrative, for since their common emergence in the sixteenth century, both the modern state and colonization posed – and continue to pose – the West the issue of the philosophical base of order' (Trouillot, 1991: 32).

figure the seamless web of technology and society for historical analysis; they also configure the Savage Slot that accommodates categories such as 'the native'. When Hughes' theory (and other theories relating to the seamless web of technology and society) is applied in empirical analyses, it continues to erase 'the native' by relegating him to the Savage Slot. In this way, he is kept offstage and does not qualify as a relevant unit of analysis for the study of electrification in STS. Seamlessness designates the assumptions of the metaphor of the seamless web of technology and society that effect this historical erasure.

Implications of Seamlessness for STS

After all, these facts were always part of the available record. That they were rarely accorded the significance they deserve suggests the existence and deployment of mechanisms of silence that make them appear less relevant than they are, even when they are known.
Trouillot, 2003: 34–5.

The discussion has shown that the terms used to frame Hughes' story of electrification in Western society are problematical. They engender a dominant narrative on technology that '[silences] the past on a world scale' by systematically establishing the seamless web of technology and society as an ideal condition to be aspired to. The assumptions underlying the metaphor of the seamless web of technology and society come to function as strategic tools to constitute society as a normative category by producing difference and contrast to 'other' populations.

Hughes' framework for studying electrification ignores the historical connections of key protagonists – Edison, Klingenberg, Rathenau and Merz – to the history of electrification outside of 'Western society'. This erasure establishes 'the continuous centrality of the Atlantic as the revolving door of major global flows' (Trouillot, 2003: 29). In this way. electrification, as a particular manifestation of the *North Atlantic universal* 'technology', contributes to the transition narrative of world history (Chakrabarty, 2008 [2000]). The transition narrative puts into operation universal claims on the historical origins and expansion of technology from North America and

Europe to the rest of the globe. The postcolonial perspective unmasks these universal claims 'as convenient fictions of the North Atlantic' (Trouillot, 2002b: 839). The seamless web of technology and society, too, qualifies as a convenient fiction of the North Atlantic. Seamlessness gives a name to the assumptions that sustain this fiction.

Hughes' model of technological change and his metaphor of the seamless web of technology and society play an important part in the intellectual history of the field of STS. He contributed new categories to describe the changing ways of thinking about technological change in the mid-1980s. Subsequent work in the tradition of SCOT, LTS and ANT adopted elements of his theory of technological change, including the idea of the seamless web of technology and society, and his work continues to be referenced by the STS community to this day. The original object of the metaphor – a means to dissolve the boundaries of conventional categories such as 'politics, technology, science, economics, society' – has been put into operation in these empirical traditions to erase categories of all kinds. In this vein, Bruno Latour, for example, has employed the seamless web metaphor to illustrate the erasure of distinctions between nature and culture, and between humans and nonhumans (Latour, 1993).

The application of the metaphor of the seamless web in the field of STS has not been investigated in closer detail. Its general operation as dissolver of boundaries between categories as hallmark of the social constructivist approach to technology in the scholarly traditions such as SCOT, LTS and ANT more broadly, however, has been subjected to critique since the early 1990s. For example, Sandra Harding has criticized ANT's erasure of categories as dismissing such basic social factors as race, class, and gender. She maintains that, by ignoring these categories, ANT is incapable of challenging the power of racism, oligarchy, patriarchy, or eurocentrism, respectively. David Bloor (1999) and Sal Restivo (2010) have objected on similar grounds, noting that ANT's vocabulary and analytical tools cannot challenge power structures, they can only describe them. Langdon Winner (1993), too, has pointed to some of the methodological problems of the social constructivist approach to technology[236] and has lamented that 'the conventional distinction between technology and society has finally broken down altogether' (Winner, 1993: 366) for some of the proponents of this tradition. In particular, he refers to the work of Michel Callon and Bruno Latour, whose actor network theory presents a 'modern world [...] composed of actor networks' that wipe out boundaries between human and nonhuman actors. He raises important questions regarding their 'methodological premise' that treats 'living persons and nonliving technological entities' equally (Winner, 1993: 366).

> Who says what are relevant social groups and social interests? What about groups that have no voice but that, nevertheless, will be affected by the results of technological change? What of groups that have been suppressed or deliberately excluded? How does one account for potentially important choices that never surface as matters for debate and choice? (Winner, 1993: 369).

Winner points out the problems of elitism and exclusion associated with this 'mode of inquiry', which displays a 'lack of and, indeed, apparent disdain for anything resembling an evaluative stance or any particular moral or political principles that might help people judge the possibilities that technologies present' (Winner, 1993:371). Finally, he criticizes the descriptive 'agnostic' position that social constructivists of technology choose 'as regards the ultimate good or ill attached to particular technical accomplishments' (Winner, 1993: 372).

> There is also no desire to weigh arguments about right and wrong involved in particular social choices in energy, transportation, weaponry, manufacturing,

---

236   Winner specifies this group of scholars: 'Among the names of those involved in this project are a number of Europeans and Americans: H. M. Collins, Trevor Pinch, Wiebe Bijker, Donald MacKenzie, Steven Woolgar, Bruno Latour, Michel Callon, Thomas Hughes, and John Law. These and other scholars of similar persuasion are now very active doing research, publishing articles, and building academic programs. They are also openly proselytizing and even self-consciously imperial in their hopes for establishing this approach. It is clear they would like to establish social constructivism as the dominant research strategy and intellectual agenda within science and technology studies for many years to come' (Winner, 1993: 364).

agriculture, computing, and the like. Even less is there any effort to evaluate patterns of life in technological societies taken as a whole. All the emphasis is focused upon specific cases and how they illuminate a standard, often repeated hypothesis, namely, that technologies are socially constructed (Winner, 1993: 373).

Michael Hard (1993), too, criticizes Pinch and Bijkers' approach as lacking a discussion of power, stratification, and hierarchy (Hard, 1993: 414).

> By presenting a view of technology in terms of functionally arranged sociotechnical systems, we will support those who benefit from harmony and cooperation and discourage those who might benefit from conflict and opposition. We might be able to reveal both unexpected and unwanted aspects of a technology, but we will remain unable to suggest an alternative vision (Hard, 1993: 413).

Many of the problems that Winner, Hard, Harding, Bloor and Restivo identified in the social construction of technology traditions of STS – mechanisms of inclusion and exclusion, its implicit evaluative stance, its untenable claim of political neutrality, and its impeding of alternative visions – concern assumptions underlying the seamless web metaphor. In a broad sense, their analyses may be considered as criticisms of the performance of the seamless web in STS's studies of technology subsequent to Hughes' *Networks of Power*. Seamlessness provides a conceptual tool to subsume and give new focused momentum to this critique.

Why is it necessary to rephrase critique raised by established STS scholars almost thirty years ago? The reason is that the problems they sought to address persist to this day and have far-reaching consequences on the intellectual advancement of the field of STS. Their arguments have not lost validity and continue to be applicable to contemporary studies of technology in the tradition of SCOT, LTS and ANT. The critical reviews of Winner, Hard, Harding, Bloor and Restivo, and others, have not prevented the problem from perpetuating in the field of STS. On the contrary, Hughes' book remains a compulsory point of reference for the study of electrification and large technological systems. The term Seamlessness, therefore, is submitted as an attempt to change the terms of the debate so as to specify and draw renewed attention to this predicament.

The implications of Seamlessness for STS, however, reach beyond Hughes' book and the study of the history of technology. Hughes' study of electrification is considered one of the founding texts of STS of the mid-1980s. For this reason, the broader intellectual environment of the time has to be taken into account when inquiring into the implications of Seamlessness for STS. Why did Hughes propose the metaphor of the seamless web of technology and society at this point in time? How did the assumptions underlying his approach to the history of technology relate to the broader academic debates in STS of the 1980s?

The intellectual roots of Hughes' notion of the seamless web of technology and society can be traced back to the tradition of the Sociology of Scientific Knowledge (SSK) of the 1970s. Hughes, trained as a historian of science, proposed his new theory of technological change in response to intellectual debates of the late 1970s and early 1980s on internal versus external histories. In particular, Hughes proposed a new approach to the history of technology that claimed to move beyond the constraints presented by the methodological approaches of internal and external histories. He presented his book on Electrification in Western Society in this spirit: 'This book is not simply a history of the external factors that shape technology, nor is it only a history of the internal dynamics of technology; it is a history of technology and society' (Hughes, 1983:2).

Hughes' study of electrification forms part of a set of new empirical studies that '[searched] for a language and for concepts to express their new understanding of technological change' (Bijker et al., 1987:13). Importantly, these studies '[borrowed] from the sociologists of knowledge' and '[urged] historians and sociologists to open the so-called black box in which the workings of technology are found' (Bijker et al., 1987:14). In proposing this link between technology and society, Hughes applied newly emerging insights from the Sociology of Scientific Knowledge (SSK) to the historical study of technology. His theory replicates for technology the argument that the content of science can be studied in social and cultural terms, an argument put forward by scholars in the SSK tradition. SSK had been developed by a group of scholars from different disciplinary backgrounds affiliated to the University of Edinburgh, who 'had set out to understand not just the organization but the content of scientific knowledge in sociological terms' (Sismondo, 2010: 47). It extended the sociology of knowledge into the arena of the 'hard sciences'. This approach came to be referred to as the Strong Programme in the SSK.

The Strong Programme's most prominent formulation was outlined by David Bloor's four tenets for the sociology of scientific knowledge in 1973. Its central tenets were that, 'in investigating the causes of beliefs, sociologists should be impartial to the truth or falsity of the beliefs, and that such beliefs should be explained symmetrically' (Bloor 1973, in Bijker et al., 1987:18). The Strong Programme viewed all knowledge claims as socially constructed and their genesis, acceptance and rejection were to be explained by the social world rather than the natural world (Bijker et al., 1987:18).

Hughes' book on electrification in Western society does not explicitly name the Strong Programme in SSK. But there is general agreement that Hughes' approach to the history of technology may be viewed as an application of the tenets of the Strong Programme in SSK. The Strong Programme, and social constructivist research approaches from these years more generally, were seen as a promising route for the history of science and technology during that time. It provided an argument 'that one can study the concept of science and technology in social and cultural terms' (Sismondo, 2010: 55). There is also general consent on the importance of the Strong Programme in SSK for the early development of the field of STS. These intellectual roots are important for our inquiry into the implications of Seamlessness for STS more broadly.

To answer the question of the implications of Seamlessness on STS more generally, we therefore can follow the conceptual connections of Hughes' seamless web with the Strong Programme in the SSK of the 1970s and early 1980s. How do Hughes' underlying assumptions on technological change relate to analytical approaches in SSK in general, and in the Strong Programme in particular?

Hughes' theory of technological change can be interpreted as an application of ideas in the Strong Programme of SSK to the history of technology. To link the findings of the last chapter to the Strong Programme in SSK we will apply Trouillot's postcolonial approach to a book that has been described as an exercise in the Strong Programme; Stephen Shapin and Simon Schaffer's renowned book *Leviathan and the Air Pump: Hobbes, Boyle and the Experimental Life* (Shapin & Schaffer, [1985] 2011).

4. Scientific Heritage as Transition Narrative

4.1 Shapin and Schaffer's Experimental Life

The book *Leviathan and the Air-Pump: Hobbes, Boyle and the Experimental Life* published in 1985 by Steven Shapin and Simon Schaffer has been widely cited in STS circles and beyond,[237] and it continues to influence scholarly debates and approaches in the field. Reflecting back, Shapin and Schaffer describe the book as 'an instantiation of a research programme in the sociology of scientific knowledge' (Shapin & Schaffer, 2011: xIi). The authors were concerned with 'a very specific passage' (Shapin & Shaffer, 2011: xxxi) in the history of early modern science. They used the classic controversy between Robert Boyle and Thomas Hobbes in seventeenth-century England to study experimental practice. The book famously begins with the following paragraph:

> The object of this book is experiment. We want to understand the nature and status of experimental practices and their intellectual products. These are the questions to which we seek answers: what is an experiment? How is an experiment performed? What are the means by which experiments can be said to produce matters of facts, and what is the relationship between experimental facts and explanatory constructs? How is a successful experiment identified, and how is success distinguished from experimental failure? Behind this series of particular questions lie more general ones: Why does one do experiments in order to arrive at scientific truth? Is experiment a privileged means of arriving at consensually agreed knowledge of nature, or are other means possible? What recommends the experimental way in science over alternatives to it? (Shapin & Schaffer, [1985] 2011: 15).

---

237 Martin et al. rank Shapin and Schaffer's book as the fifth most cited contribution listed by STS handbook authors and refer to it as 'core literature' of STS (Martin et al., 2012).

Shapin and Schaffer's prominent book on the emergence of specific practices that came to shape early modern science offers excellent foil to probe the question of the global travel of STS and its Collective Blind Spot. The previous chapter investigated these travels by examining an influential book in the intellectual tradition of STS dedicated to the study of *technology*. This chapter explores a book published around the same time that shaped the intellectual tradition of STS concerned with the study of *science*. It picks up on the conclusions on technology as a *North Atlantic universal*, which demonstrated Michel-Ralph Trouillot's claim that these words carry a vision of the world that requires an alterity – a constitutive Otherness. Expanding on this conclusion, the chapter starts out with the claim that science, in its guise as *North Atlantic universal*, also requires an alterity against which its knowledge claims attain their full meaning. To demonstrate this claim, Trouillot's concept of alterity ('Elsewhere') is applied to Steven Shapin and Simon Schaffer's paradigmatic book.

If the hypothesis is correct, the case study will have to demonstrate that Shapin and Schaffer's historical analysis of early modern science requires a relation to Otherness. An analysis of Trouillot's geographies of management and imagination as they relate to this book should reveal the inhabitants of its Elsewhere, the oppositional referents for casting and legitimizing their knowledge claims. The chapter will first offer a brief account of Shapin and Schaffer's book on Experimental Life. Then, the philosophical programmes of Hobbes and Boyle are examined through the lens of the geographies of management and imagination to show that they required a relation to Otherness. The 'New World', 'savages' and 'inferior creatures' figured as oppositional referents for casting and legitimizing their knowledge claims. Elsewhere is shown to reveal a mechanism by which early modern science comes to gain legitimacy. The key questions and answers of Shapin and Schaffer's book are then revisited against this background. This examination shows that Shapin and Schaffer required the residual category of the 'ignorant stranger' as a crucial referent for framing their symmetrical historical approach to experiment.

An Exercise in the Sociology of Scientific Knowledge

Shapin and Schaffer's book deals 'with the historical circumstances in which experiment as a systematic means of generating natural knowledge arose, in which experimental practices became institutionalized, and in which experimentally produced matters of fact were made into the foundations of what counted as proper scientific knowledge' (Shapin & Schaffer, [1985] 2011: 3). The authors proposed a new approach to studying one of the 'great paradigms' of historians of science, who regarded Robert Boyle as 'a founder of the experimental world in which scientists now live and operate' (Shapin & Schaffer, [1985] 2011: 5).

The study by Shapin and Schaffer departs from the postulate in the history of science that the historian and the 17th-century experimentalist Robert Boyle share a culture. Although the authors do not challenge this postulate, they do challenge the historical 'member's account and its associated self-evident method' (Shapin & Schaffer, [1985] 2011: 5) that historians of science usually produced. Instead, they propose a new approach to studying experiment, that of 'playing the stranger' to 'the culture of experiment' (Shapin & Shaffer, [1985] 2011: 6). Their key methodological challenge, therefore, is to specify how the historian can 'play the stranger to experimental culture, a culture we are said to share with a setting in the past and of which one of our subjects is said to be the founder?' (Shapin & Shaffer, [1985] 2011: 6). Despite this shift of focus to 'culture', Shapin and Schaffer emphasize their positioning in history rather than anthropology and present their book as 'an exercise in the sociology of scientific knowledge' (Shapin & Schaffer, [1985] 2011: 15).

The author's historical approach to experimental culture is to cast the 'great paradigm of experimental procedure' in terms of a controversy within the social context of 17th-century Restoration England. This controversy took place between two principal protagonists – Robert Boyle and Thomas Hobbes – in England in the 1660s and early 1670s (Shapin & Schaffer, [1985] 2011: 7). Shapin and Schaffer prefer a symmetrical handling of rejected and accepted knowledge over controversy to rectify what they view as an asymmetrical, standard historiographical strategy of naming Boyle as the winner of the Boyle–Hobbes controversy (Shapin & Schaffer, [1985] 2011: 11). For this purpose, Shapin and Schaffer attempt a member's account of Hobbes's 'anti experimentalism' (Shapin & Schaffer, [1985] 2011: 13) and a stranger's account of Boyle's experimental programme. In this way, Shapin and Schaffer evoke three positions of historical knowing: the historian as a member of experimental culture, the historian as an artificial (playing, pretending) stranger to experimental culture, and the (non-) historian as a genuine (ignorant) stranger to experimental culture.

The central argument of the book is that the Hobbes–Boyle controversies show that '[...] solutions to the problem of knowledge are embedded within practical solutions to the problem of social order, and that different practical solutions to the problem of social order encapsulate contrasting practical solutions to the problem of knowledge' (Shapin & Schaffer, [1985] 2011: 15). Shapin and Schaffer conclude that the Restoration polity and experimental science shared a common form of life and that 'the practices involved in the generation and justification of proper knowledge were part of the settlement and protection of a certain kind of social order' (Shapin & Schaffer, [1985] 2011: 342).

The New World, 'Inferior Creatures' and 'Savages'

Shapin and Schaffer conclude that both Hobbes and Boyle's philosophical programmes shaped the nature of early modern philosophical life and experimental culture. This chapter claims that their philosophical programmes required an alterity, a referent outside themselves, a pre- or non-science in relation to which their programmes attained their full meaning. What alterities do Hobbes and Boyle's philosophical programmes require? What is the complementary Elsewhere to the Here of their philosophical programmes that both premised one another and were conceived as inseparable? What Others inhabit the Elsewhere in Hobbes and Boyle's philosophical programmes?

Trouillot gives a name to the Elsewhere that was required for the emergence of modern Europe. He describes the 'discovery' of the New World, America, as the creation of Europe's 'still unpolished alter ego, its elsewhere, its Other' (Trouillot, 1991: 23). Indeed, Trouillot considers 'The Conquest of America [...] as Europe's model for the constitution of the Other' (Trouillot, 1991: 23). Stuart Hall (1992) identifies an explicit connection between the quest for order in Europe and the construction of a modern identity. According to Hall, this identity was formed not only by 'the internal processes that gradually moulded Western European countries into a distinct type of society but also through Europe's sense of difference from other worlds—how it came to represent itself in relation to these "others"' (Hall, 1992: 189). Hall's approach corresponds to Trouillot's views on 'the West's vision of order', which 'from its inception required two complementary spaces, the here and the elsewhere, which premised one another and were conceived as inseparable' (Trouillot, 2002a: 222).

Because Shapin and Schaffer's book is concerned with the emergence of early modern science in 17th-century England, Trouillot's specification of Elsewhere for early modernity may be transferred to this case: Elsewhere is the New World, America. Although Shapin and Schaffer do not mention the New World, or its inhabitants in their book, other researchers have studied Hobbes and Boyle's connections with the New World and their images of its inhabitants. They have shown that Hobbes and Boyle used the expressions 'savage' (Moloney, 2011) and 'inferior creatures' (Irving, 2008), respectively, to designate the populations of the New World in their work.

4.2 Hobbes's and Boyle's Natural Philosophical Programmes

Elsewhere in Robert Boyle's Natural Philosophical Programme

Robert Boyle never travelled to the New World. However, he held influential positions in various colonial institutions in the New World: the English East India Company, the Council for Foreign Plantations and the New England Company (Irving, 2008: 1).[238] He also held stocks in the Hudson's Bay Company (Irving, 2008: 1). Although he never visited America, Boyle actively entertained contact with travellers, tradesmen and colonial officers who provided him with first-hand information about the New World. In the absence

---

[238]   Irving writes that 'Boyle served on the board of the English East India Company and on the Council for Foreign Plantations, acted as President of the New England Company and held shares in the Hudson's Bay Company' (Irving, 2008: 1).

of established conventions, codes and practices for validating knowledge in seventeenth-century England the verification of information from the New World in 17th-century England was a serious practical problem to natural philosophers. Boyle solved this problem by establishing a network of informants, a community of messengers or harbingers, who provided him with hands-on information about the New World. Irving shows that the 'gentlemanly reliability' of these informants was established by considering these 'men's associations with England's colonies' (Irving, 2008: 70).[239]

Shapin and Schaffer show that Boyle's membership in the Hartlib Circle and the Royal Society influenced his philosophical programme. Thus, their analytical focus remains on Boyle's 'elaboration and implementation of procedures and institutions of control' (Trouillot, 2003: 2) *at home*. They do not consider the influence of Boyle's association with procedures and institutions of control *abroad*. However, both the Royal Society and the Hartlib Circle viewed the New World as a source of useful knowledge for natural philosophers and they established overseas channels of communication accordingly (Irving, 2008: 69).

Boyle's primary motivation, however, was not to establish new commonwealths in the New World, to derive profit or to gain first-hand knowledge from informants (Irving, 2008). Managing these interests served his primary goal, the re-creation of 'The Empire of Man over Inferior Creatures' (Irving, 2008: 1).[240] In his tract, *Of the Usefulness of Experimental Natural Philosophy*, Boyle considers travellers to be vital to 'the natural philosopher's epistemic project of creating man's empire of knowledge' (Irving, 2008: 14).[241] Boyle believed that 'only experimental philosophy could give him

---

239   Irving bases this claim on her analysis of the work diaries of Boyle (in which he reports on interviews with travellers) by focusing on 'the way he dealt with the epistemological issue of the veracity of information collected on the colonial periphery' (Irving, 2008: 70).

240   R. Boyle, Some Considerations Touching the Usefulness of Experimental Natural Philosophy, Part II (1671), in M. H. Hunter and E. B. Davis (eds.), The Works of Robert Boyle, 14 vols. (London: Pickering and Chatto 2000), vol. 6, p. 406 (Irving, 2008).

241   See also Irving, 2008: 69.

access to the New World' (Irving, 2008: 74).[242] Furthermore, Irving shows that Boyle considered the recovery of man's original empire as 'part of a programme for a model life as an experimental philosopher and as a Christian' (Irving, 2008: 77). His model for the natural philosopher was 'the Christian Virtuoso' (Irving, 2008: 78).[243]

Because Boyle never travelled to the New World, and because there were no established procedures to verify knowledge at the time, Boyle's expression 'inferior creatures' must be classified as a figment of his imagination. The role of 'inferior creatures' in his philosophical programme and the space this expression occupied in his claims for knowledge, however, are not imaginary. Its assimilation into his body of knowledge is what Pratt referred to as a normalizing discourse that codifies difference by omitting the 'observing self' in its textual production of the Other (Pratt, 1992: 140).

Mapping and relating the geography of management and imagination of Boyle's philosophical programme reveals the direct connections and entanglement between scientific institutions such as the Hartlib Circle and the Royal Society, and colonial institutions such as the English East India Company, the Council for Foreign Plantations and the New England Company. The specific geography of Boyle's imagination required finding a slot for the inhabitants of the New World, inferior creatures, and a model of their absolute opposite, the experimental natural philosopher.

---

242 'In his defence of experimental philosophy, Boyle drew a theological connection between natural philosophers' pursuit of man's original empire over the world, and English trade and colonization, which he argued were the means of fulfilling God's command to man to enjoy the fruitfulness of the earth' (Irving, 2008: 69).

243 'The reason why English natural philosophers were so interested in the New World, and particularly in English colonies, was because they held a theory of empire as man's original dominion over nature. It was the emphasis upon restoring man's original encyclopaedic knowledge of the natural world which generated their interest in the New World, and ultimately in its English colonies' (Irving, 2008: 3).

Elsewhere in Thomas Hobbes's Philosophical Programme

Just like Boyle, Hobbes never visited the New World, and just like Boyle, Hobbes was directly involved with colonial institutions in the New World. As secretary to Lord Cavendish, Hobbes was professionally acquainted with the New World through their 'joint involvement with the Virginia Company' (Moloney, 2011: 197).[244] The Virginia Company was a joint stock company whose purpose was to establish colonial settlements in North America. Its governing activities were carried out 'according to a series of royal charters' (Rose, 2001[245]).

There are few historical records on Hobbes's involvement in the Virginia Company. Indeed, Malcolm considers Hobbes to have remained silent about his own place in the story of the Virginia Company (Malcolm, 1981: 301). At the same time, he believes that Hobbes 'made far more use of the New World than was previously recognized' (Moloney, 2011: 197). Hobbes's direct connection with colonial affairs gave him access to 'detailed reports of the Amerindians' (Moloney, 2011: 197). He used the New World as a crucial referent in his seminal work on political sovereignty, in which he built an account 'of the sovereign authority of civilized states that presupposed its absence among the primitive societies of the New World' (Moloney, 2011: 199). Moloney describes how Hobbes only saw 'savage anarchy' in the New World and contrasted savage societies of the Americas with European states: 'Primitive societies were not considered to be political units whose sovereignty merited respect. These were instead the outside of the system of sovereign states; the chaotic and barbaric regions over which civilized states

---

244   Moloney quotes the work of Aravamudan (2009) and Malcolm (2002) who have considered Hobbes's involvement in English colonial America. Hobbes was employed as a tutor to Lord Cavendish (who was only two years younger than Hobbes) after leaving Oxford. When Lord Cavendish started a political career at court and in parliament, 'Hobbes' employment in the Cavendish household gradually changed from that of a tutor to that of a secretary'. For several years, the 'most important and time-consuming business interest of [Hobbes'] pupil-patron' was the Virginia Company' (Malcolm, 1981: 297).

245   Hobbes too gained membership of the Company, and there are indications that he was a shareholder in the Somer Islands Company, an 'independent but largely subsidiary company responsible for the settlement of the Bermudas' (Malcolm, 1981: 298).

competed in the extension of rival imperial jurisdictions' (Moloney, 2011: 199).

On what basis did Hobbes come to see savage anarchy, chaos and barbarism in the New World, even though he never travelled there? Hobbes's association with colonial institutions gave him access to current news and descriptions from travellers, tradesmen and colonial officers. Hobbes therefore relied on the same type of information on the New World and its inhabitants as Boyle did, which may be studied as a normalizing discourse that textually produces the Other (Pratt, 1992). Hobbes used the New World as a crucial referent in his seminal work on political sovereignty. Hobbes made an analogy between sovereign individuals in a state of nature and sovereign states in the European family of nations. This analogy necessitated the denial of statehood to savage societies. Hobbes denied the indigenous societies of the New World statehood by identifying them as states of nature. He then conceptualized 'the sovereign, European state in juxtaposition to a hypothetical state of nature' (Moloney, 2011: 199). Assigning 'savage societies' to nature meant to classify their social structures and institutions as pre-political (Moloney, 2011).

In sum, Hobbes built an account 'of the sovereign authority of civilized states that presupposed its absence among the primitive societies of the New World' (Moloney, 2011: 199). His analogy between sovereign individuals in a state of nature and sovereign states in the European family of nations necessitated the denial of statehood to savage societies (Moloney, 2011: 199). Assigning 'savage societies' to nature meant to classify their social structures and institutions as pre-political (Moloney, 2011). Moloney points out that this kind of theoretical mapping fed into 'the ideologies Europeans used to rationalize the global order they were bringing into being' (Moloney, 2011: 190). He writes that 'Hobbes built a theoretical bridge between the chaos of the colonial periphery and the order that ought to characterize the internal structures of, and the external relationships among, European states' (Moloney, 2011: 190).

By inscribing 'savages' as inhabitants of 'the anarchic periphery' of the New World, Hobbes assigned them a backward space in both time (before sovereign states) and place (outside sovereign states). By theorising 'savage anarchy [...] as the only alternative to, the very outside of, civilized states' (Moloney, 2011: 202), Hobbes contributed to the development of those binary oppositional concepts that have sustained the idea of a European modernity. Instead of viewing these binary constructions as agents in developing modernity as a stage in history, they can be seen as casting modernity as historiographical frame for staging difference (Mitchell, 2000).

Mapping and relating the geography of management and imagination of Hobbes's philosophical programme reveals that Hobbes's involvement and connections to colonial institutions in the New World are meaningful for his philosophical programme. Hobbes's theory necessitated what Trouillot has termed 'the Savage Slot' (Trouillot, 2003), an Otherness against which core concepts of his philosophical programme (and their superiority) came to make sense in 17th-century Restoration England. Hobbes's 'savage' is not just a figment of the imagination that he projected into the New World that would correspond more or less to a reality. The use of the 'savage' in his philosophical programme becomes what Homi Bhabha has referred to as a 'strategy of representing authority' (Bhabha, 1994). This strategy assigned backwardness in time and instantiated an order (of sovereign citizenship) against which variations can be discussed.

4.3 "What is Experiment?"

The above analysis shows that Boyle and Hobbes's philosophical programmes necessitated an Elsewhere as referent against which their knowledge claims could attain full meaning and legitimacy. But it is not clear whether this approach offers new answers to Shapin and Schaffer's principal research questions (What is experiment? What were the historical circumstances in which modern experimental life became institutionalized?). Shapin and Schaffer do not define experiment. Instead, they describe characteristics that typify the phenomenon of experimental practice: experiment is *a solution* (to the problems of knowledge and social order); experiment is *a cultural practice* (brought about through consensus produced by historical judgement); and experiment is *a form of life* (shared by the politics and science of the Restoration society, which judges what will best establish order). The following sections will present new insights from this study on Elsewhere with respect to these characteristics.

Experiment as Solution to Knowledge and Order

The analysis has shown that Hobbes and Boyle used the expressions 'the savage', 'wilderness' and 'inferior creatures' to populate Elsewhere in their respective philosophical programmes. At the most basic level, the analysis suggests an essential complement to Shapin and Schaffer's central tenet of viewing Hobbes and Boyle's programmes as solutions to the problems of knowledge and social order. Experimental practice as a solution to the problems of knowledge and social order is at the same time a solution to the problems of *non-knowledge* and social *disorderliness*. As will be discussed later, this is not merely another exercise in symmetry.

Experiment as Cultural Practice

Shapin and Schaffer are interested in explaining the beliefs and practices of experimental culture (Shapin & Schaffer [1985] 2011). Representing the beliefs and practices of a specific culture sets the boundaries for its cultural identity. According to Corbey and Leerssen (1991b: vi), 'the circumscription of cultural identity proceeds by silhouetting it against a contrastive background of Otherness'. Van Alphen shows that descriptions of alterity '[…] are never based on a 'real' Other, but on a denial of the self, of the observer's identity' (van Alphen, 1991: 3).

Thus, examining experiment as culture and practice means representing cultural identity, and the portrayal of cultural identity requires a contrastive background of Otherness that cannot be based upon a 'real' Other but must be based on an imaginary projection that constitutes the self. Hobbes and Boyle used 'savages' and 'inferior creatures' as means to formulate '[…] an ideal, desired identity' (van Alphen, 1991: 3): the civilized (European) state and the model experimental philosopher. To view experiment as culture, therefore, not only implies the need to connect what happens inside the experimental laboratory with social conventions and formative socio-historical contexts; it also entails a shift towards the geography of imagination that evokes 'the twin concepts of alterity and identity' (Voestermans, 1991: 221).

Why is this difference important? Shapin and Schaffer regard *the process of social consensus* on Boyle and Hobbes's knowledge claims as the driving force in producing historical judgement in favour of experimental life. This study considers the *construction of a modern European identity* as the key to the successful insertion of both Hobbes and Boyle's programmes into the specific circumstances of Restoration England. Experimental practice becomes a device to construct meaning, a site of enunciation to inscribe difference and sameness.

Experiment as a Form of Life

Shapin and Shaffer depict experiment as a form of life. In their view, the ability of this form of life to establish social order was judged by the context of Restoration society. Boyle's experimental form of life, however, was not proposed simply as an optional form of life among others but as a *model* form of life, the 'Christian Virtuoso' (Irving, 2008). Blending the figure of the experimental philosopher with the Christian composed a model identity that allowed Boyle to bestow his programme with a sense of social, political, cultural, economic and religious utility.

Furthermore, Boyle's model form of life was cast against the contrasting forms of life of 'inferior creatures'.[246] Its aim was to recover 'man's knowledge of the natural world' to 'exert dominion over the Creation' (Irving, 2008: 78). The order sought by Hobbes and Boyle's philosophical programmes concerned not only the battles of the English nation outside the confines of the 17th-century laboratory, but also its (Christian) handling of the encounter with the New World. The idea that 'Dominion and knowledge [...] go hand in hand' (Irving, 2008: 78)[247] was shared by natural philosophers, especially the members of the Royal Society, at the time. This blending can be seen as

---

[246]   Likewise, Hobbes's philosophical programme necessitated 'the savage' in the wilderness of the New World, as a referent against which to inscribe his concept of political sovereignty. His model of civilized states both presupposed and required the denial of statehood to savage societies.

[247]   'In Boyle's work the idea of man's empire became a detailed theory in which he explored the nature of empire as a form of power: its origins, theological legitimacy and finally its role as a pursuit for the model natural philosopher, the Christian Virtuoso' (Irving, 2008: 78).

an 'attempt to subdue the strangeness of the Other in cognitive terms' (Corbey & Leerssen, 1991b: viii). Accordingly, the model experimental life was developed in the context of the construction of the identity of Western European nations vis-à-vis the New World – and their struggle for colonial power. Its particular social imaginary (Appadurai 1999), therefore, formed part of the larger process of inventing early modern selves of modernity. Randeria has referred to this identity as the precursor stage of the European Self ('*Vorstufe des europäischen Selbst*') (Randeria, 1999b: 374). To produce the coherence and sameness of this identity, Boyle and Hobbes's philosophical programmes needed to project alterity and tame difference.

One might be tempted to conclude that these new answers to Shapin and Schaffer's questions simply require Shapin and Schaffer's historical record to be complemented by extending the context of Restoration society to other geographical regions, such as the New World, or to other cultures, designated wrongly as 'savages' or 'inferior creatures'. Is this really all that happens, however, if we acknowledge that modern science requires 'an alterity, a referent outside itself, a pre- or non-modern' in relation to which modern science attains its full meaning? A shift takes place that moves beyond focusing on linear processes of inclusion or travel to focusing on processes of co-constitution and entanglement (Randeria, 2002). The replication of these alterities does not simply happen by overseeing historical realities that have now been uncovered by a newer generation of scholars, such as Moloney, Irving, and others. The replication of alterities lies concealed in Shapin and Schaffer's particular approach to experiment: by viewing and treating experiment as a cultural heritage of modern science.

Experiment as Heritage

In addition to these explicit characteristics, Shapin and Schaffer's analytical approach implies a significant assumption that informs their approach to experiment: their distinction between member's and stranger's accounts assumes that Boyle's experimental programme shaped a cultural practice that has a *historical lineage* reaching back to its *origins* in 17th-century England.

Experimental practice is thus represented as tradition or *cultural heritage*.[248] Shapin and Schaffer re-inscribe the image of early modern experimental practice as a culture with directional historical footprints that can be extrapolated and traced through time and across space.[249] This image provides the backdrop to their proposition for a new approach to the question: 'How can the historian play the stranger to experimental culture, a culture we are said to share with a setting in the past and of which one of our subjects is said to be the founder?' (Shapin & Schaffer, [1985] 2011: 6). Shapin and Schaffer propose the research methodology of 'playing the stranger' to 'the culture of experiment' (Shapin & Schaffer, [1985] 2011: 6):

> We need to *play* the stranger, not to *be* the stranger. A genuine stranger is simply ignorant. We wish to adopt a calculated and an informed suspension of our taken-for-granted perceptions of experimental practice and its products. By playing the stranger, we hope to move away from self-evidence. We want to approach "our" culture of experiment as Alfred Schutz suggests a stranger approaches an alien society, [...] (Shapin & Schaffer, [1985] 2011: 6).

This methodology presupposes that the philosophical programmes of both Boyle and Hobbes form part of the cultural heritage of experimental practice (because each was offered as a solution to the problems of knowledge and social order) and that these programmes can be studied by applying a

---

248   'Thus, historians start with the assumption that they (and modern scientists) share a culture with Robert Boyle, and treat their subject accordingly: the historian and the seventeenth century experimentalist are both members' (Shapin & Schaffer [1985] 2011: 5).

249   Shapin and Schaffer position their approach in contrast to the conventional historian's approach to experiment. They challenge the historian's 'self-evident method' of constructing a member's account (Shapin & Schaffer, [1985] 2011: 5). 'We have a dismissal, the rudiments of a causal explanation of the rejected knowledge (which implicitly acts to justify the dismissal), and an asymmetrical handling of rejected and accepted knowledge' (Shapin & Schaffer, [1985] 2011: 11). 'One reason why historians have not systematically and searchingly pressed the questions we want to ask about experimental practices is that they have, to a great extent, been producing accounts coloured by the member's self-evident method. In this method the presuppositions of our own culture's routine practices are not regarded as problematic and in need of explanation' (Shapin & Schaffer, [1985] 2011: 5).

methodology from the Strong Programme in the Sociology of Scientific Knowledge (Bloor, 1991 [1976]): the principles of symmetry and impartiality. By treating experiment as a cultural heritage shared by the historian of science, the 17th-century experimentalist, and Boyle, Shapin and Schaffer replicate the alterities that constitute modern science. Who is the genuine stranger in Shapin and Schaffer's approach to experimental culture? What is he/she ignorant of? In what sense is he/she not believable?[250] What is inherent in his/her views that allows the two STS scholars Shapin and Schaffer to evaluate and assign to him/her the status of the ignorant?[251]

Because Shapin and Schaffer do not elaborate further on the genuine, ignorant stranger, we can only speculate about his/her identity. A few characteristics, however, are evident: the genuine, ignorant stranger designates the opposite, the negative necessary to frame their approach to experimental culture. The genuine, ignorant stranger, therefore, is a device to construct meaning. He/she is not based on any empirical, analytical or theoretical

---

[250] Shapin and Schaffer's methodology rests upon the assumption that both Hobbes and Boyle's philosophical programmes were believable: 'Given other circumstances bearing upon that philosophical community, Hobbes's views might well have found a different reception. They were not widely credited or believed—but they were believable; they were not counted to be correct, but there was nothing inherent in them that prevented a different evaluation' (Shapin & Schaffer, [1985] 2011: 13). This claim is contrasted against the genuine ignorant stranger, who by implication is not believable and carries something inherent in him that prevents a different evaluation.

[251] 'As part of the same exercise we shall be adopting something close to a "member's account" of Hobbes's anti-experimentalism. That is to say, we want to put ourselves in a position where objections to the experimental programme seem plausible, sensible, and rational. Following Gellner, we shall be offering a "charitable interpretation" of Hobbes's point of view. Our purpose is not to take Hobbes's side, nor even to resuscitate his scientific reputation (though this, in our opinion, has been seriously undervalued). Our goal is to break down the aura of self-evidence surrounding the experimental way of producing knowledge, and "charitable interpretation" of the opposition to experimentalism is a valuable means of accomplishing this. Of course, our ambition is not to rewrite the clear judgement of history [...]. Yet we want to show that there was nothing self-evident or inevitable about the series of historical judgements in that context which yielded a natural philosophical consensus in favour of the experimental programme' (Shapin & Schaffer, [1985] 2011: 13).

studies on the nature of his/her ignorance or lack of credibility. Therefore, he/she has to be classified as a fiction of Shapin and Schaffer's imaginary geography of experimental culture. This 'complete stranger' to experimental culture is the necessary device that allows Shapin and Schaffer to speak about 'culture': if he/she did not exist (in their geography of imagination), there would be no outside of experimental culture, and the very notion of 'culture' would lose its meaning.

Shapin and Schaffer apply the principle of symmetry to two types of knowledge: rejected and accepted. The analytical web of member's and (portrayed) stranger's accounts, however, eclipses the historical complementarity of representing knowledge against its opposite, that is, non-knowledge. To eclipse this dimension is to replicate the alterities that constitute experimental life and thus modern science more generally. This is not just another level of symmetry: the added value of making visible the alterities of modern science lies not in the claim for an equal treatment but in the claim for a need to focus on the ways in which the twin pair of identity and alterity comes into being, functions, interacts, and changes.

This approach may be viewed as an example of what Dipesh Chakrabarty has termed the 'transition narrative' of historicism (Chakrabarty, 2008 [2000]). It expresses a 'historical construction of temporality' in which the modern is separated from the pre-modern by historical time (Chakrabarty, 1992: 13). It also records the assumed cultural distance between the West and the non-West. This 'mode of thinking about history' assumes that 'any object under investigation [retains] a unity of conception throughout its existence' (Chakrabarty, 2008 [2000]: xiv). The transition narrative inscribes a "'first in Europe, then elsewhere" structure of global historical time' (Chakrabarty, 2008 [2000]: 7) that situates the modern individual at the very end of history (Chakrabarty, 2008 [2000]: 10). The mode of writing history as a transition narrative, in Chakrabarty's theory, represents a rehearsal of the split between the modern and the pre- or non-modern: '[...] this split is what is history; writing history is performing this split over and over again' (Chakrabarty, 1992: 13). Performing this split relegates the colonial subject to an 'Imaginary Waiting Room of history' (Chakrabarty, 2008 [2000]: 8).

The examination of experimental culture as heritage may be viewed as a rehearsal of Chakrabarty's transition narrative, which relegates the ignorant stranger to experimental culture to the Imaginary Waiting Room of history.

Shapin and Schaffer's book re-enacts a social imaginary in which experimental practice as culture is cast against an outside referent (Elsewhere). To view experiment as a cultural heritage means to replicate its constitutive alterities, to extrapolate and re-project them through time, all the way into the present – to perform the split of historical writing over and over again. Just as Randeria posits for modernity (Randeria, 2002), the idea of modern science travelling to the rest of the world must be replaced by a more messy and complex picture. This picture might be referred to as entangled experimental culture, or as uneven modern sciences.

The above discussion shows that the analysis of Elsewhere in Hobbes and Boyle's philosophical programmes offers new insights into the central questions of Shapin and Schaffer's book. STS appears to reproduce the alterities that have constituted modern science by replicating Chakrabarty's 'first in Europe, then Elsewhere' structure of global historical time (Chakrabarty, 2008 [2000]: 7) and by inscribing this time as a measure of cultural distance. This performance of Chakrabarty's historical split within the body of STS presents an obstacle to the advancement of this field, because it prevents the genuine, ignorant stranger from STS's own Elsewhere from entering this field of knowledge.

Shapin and Schaffer used the classic controversy between Robert Boyle and Thomas Hobbes in seventeenth-century England to study the broader questions, 'What did people actually do when they were making what they considered to be knowledge? How did they warrant what they produced, and how did they secure credibility and authority for it?' (Shapin & Schaffer, 2011: xiviii). The analysis of Elsewhere in the philosophical programmes of Boyle and Hobbes has confirmed the hypothesis that modern science requires a referent outside itself, a pre- or non-modern science in relation to which it attains its full meaning. Shapin and Schaffer's approach to experiment as culture and practice in *Leviathan and the Air-Pump* also relies on the oppositional contrast between member's and stranger's accounts. Their symmetrical account of the Hobbes–Boyle controversy in terms of rejected and accepted knowledge puts into effect the residual category of the 'genuine', 'ignorant stranger', the outside to the culture of experiment. The genuine, ignorant stranger represents a rhetorical device for legitimizing their historical method and serves as a critical referent in relation to which their historical knowledge claims attain full meaning. He/she is an imaginary construct, an inhabitant of *Leviathan and the Air-Pump*'s Elsewhere. This appropriation of a heritage of experimental culture (by contrasting it with its absence, Elsewhere) promulgates the grand narrative of modernity in which experimental life appears with a European site of origin spreads over time and place. Shapin and Schaffer's study thus replicates the alterities that have constituted modern science.

5. A Postcolonial Programme in STS

Can general implications be drawn from the postcolonial analysis of the paradigmatic books by Thomas Hughes, and Steven Shapin and Simon Schaffer? Both books perform elements of the Strong Programme in the Sociology of Scientific Knowledge (SSK). *Networks of Power* applies principles from the Strong Programme to the history of technology, and *Leviathan and the Air Pump* has been called one of its most celebrated products (Longino, 2002). For this reason, the four tenets of David Bloor's version of the Strong Programme will be used as referents for formulating a Programme in Science Studies Elsewhere (Bloor, [1976] 1991). This programme employs Shalini Randeria's concepts of *entangled histories* and *uneven modernities* (Randeria, 1999a, 1999b) to put into effect and interpret Michel-Rolph Trouillot's notion of *Elsewhere* for the study of modern science and technology. To formulate the Programme in Science Studies Elsewhere (SSE), this chapter first proposes the notion of *Science & Technology Studies Elsewhere*[252] and introduces Randeria's concepts of entangled histories and uneven modernities.

5.1 Science & Technology Studies Elsewhere

The study of alterity and the Other may be encountered in a variety of scholarly and disciplinary approaches. *Elsewhere* in this book refers specifically to Trouillot's study of alterity and the Other. Accordingly, the notion of Science & Technology Studies Elsewhere does not simply imply STS in far-away

---

[252] For the sake of brevity, the *Programme in Science Studies Elsewhere* leaves out the word 'technology' as contained in the expression *Science & Technology Studies Elsewhere*. Technology is subsumed under the heading of the programme.

places, nor is it merely a call to include peripheral voices to achieve a more representative cultural composition of STS scholarship. Michel-Rolph Trouillot's notion of Elsewhere refers exclusively to the representation of alterity in the project of modernity and emphasizes imagination and space (rather than place). In his view, modernity 'requires an alterity, a referent outside of itself—a pre- or non-modern in relation to which the modern takes its full meaning' (Trouillot, 2002a: 222).

> The claim that someone—someone else—is modern is structurally and necessarily a discourse on the Other, since the intelligibility of that position—what it means to be modern—requires a relation to Otherness. The modern is that subject which measures any distance from itself and redeploys it against an unlimited space of imagination (Trouillot, 2002a: 226).

Science & Technology Studies Elsewhere adopts Trouillot's premise for modernity and claims that modern science and technology, too, require an alterity, a referent outside itself in relation to which they take their full meaning. The phrase casts science and technology as *North Atlantic universals* which can be studied through the lenses of the geographies of management and imagination. The objective of Science & Technology Studies Elsewhere is to capture and make visible the transition narrative of science and technology. The transition narrative refers to a 'historical construction of temporality' in which the modern is separated from the pre-modern by historical time (Chakrabarty, 1992: 13).

The analysis has shown that Trouillot's notion of Elsewhere can call attention to characteristics of science and technology that have remained systematically eclipsed and that it reveals important sites for future STS. This field has assumed a particular 'gaze' on science and technology, rooted in this field's North Atlantic socio-political, institutional and epistemic heritage, and its positioning within the paradigm of modernity. If science and technology require an alterity, a referent outside itself, and if STS fails to demonstrate that it has considered this alterity in its study of science and technology, then its 'gaze' is likely to have led to a re-inscription of the alterities that have been constitutive of science and technology. The key, then, is not to seek ways to return this gaze by pluralizing the set of authors, problems or empirical sites in the field of STS. It becomes necessary to analyse the Elsewhere, the space 'of and for the Other' that can be, and often is, imaginary (Trouillot, 2002a: 224) in science and technology, and in STS. The term Science & Technology Studies Elsewhere is coined to designate this analytical focus.

5.2 Entangled Histories and Uneven Modernities

The construction of alterity and the staging of differences are matters of central concern to the postcolonial conception of modernity. Trouillot's notion of Elsewhere belongs to a particular postcolonial conception of modernity, which essentially differs from the assumptions on modernity that underlie the works by Bloor, Hughes, Shapin and Schaffer in the 1970s and 1980s. In the following decade, the idea of modernity and its underlying images of Western progress and development were increasingly criticized. By the turn of the century, concepts such as *alternative, multiple* and *plural modernities*[253] had been proposed to overcome the confines of a singular conception of modernity. Modernity is a constitutive category of thought in the social sciences and humanities. It is also a constitutive category of thought in the field of STS. Therefore, developments in the conception of modernity have a bearing on the work of STS.

Modernity has become 'a contested concept with a multiplicity of meanings which vary with actors and contexts' (Randeria, 2002, 287). Classically, modernity was portrayed as a result of the economic, socio-political and cultural movements connected to the European Renaissance, the Reformation and the Scientific Revolution. These European developments were seen as a blueprint for modernity to be followed by the rest of the (non-modern) world. Mitchell (2000) describes the modern age as presenting 'a particular view of geography, in which the world has a single center, Europe [...] in reference to which all other regions are to be located; and an understanding of history in which there is only one unfolding of time, the history of the West, in reference to which all other histories must establish their significance and receive their meaning. [...] Historical time, the time of the West, is what gives modern geography its order, an order centered upon Europe' (Mitchell, 2000: 7).

---

[253] See Knauft (2002) for an overview of relevant literature on these concepts.

The concepts of alternative, multiple and plural modernities were proposed to transcend the Eurocentric view of historical modernity. These concepts, in turn, have been criticized by scholars in postcolonial studies, cultural studies and anthropology. According to Conrad and Randeria (2002: 10), alternative or parallel forms of modernity reproduce the boundaries of the nation-state and Europe. Bhambra writes that 'simply pluralizing the cultural forms of modernity, or recognizing the histories of others, does nothing to address the fundamental problems with the conceptualization of modernity itself' (Bhambra, 2011: 655). On the contrary, such approaches continue to re-inscribe the very categories and polarities that they seek to overcome. Modernity thus continues to provide what has been referred to as a 'historiographical frame' (Mitchell, 2000) or 'grand narrative, within which the origin and diffusion of modernity within Europe is located' (Bhambra, 2011: 653). Randeria criticizes this grand narrative for casting world history in terms of binary contrasts, which perpetuates the view of 'European historical experience [...] as both unique and universal' (Randeria, 2002: 291).

The most recent postcolonial approaches to modernity have been concerned with finding ways to move beyond the conception of modernity as a stage in history. Rather, modernity is viewed as the 'staging' of history (Mitchell, 2000: 23), a process that 'involves the staging of differences' (Mitchell, 2000: 26). Often, this act is referred to as the construction of alterity. The requirement of constructing alterity for the genealogy of modernity has become the focal point of recent analysis in a variety of academic fields, such as postcolonial studies, anthropology, cultural studies and subaltern studies. Important contributions to this debate have been made by (among others) Edward Said (1977), Stuart Hall (1992, 1997), Homi Bhabha (1994), Gayatri Chakravorty Spivak (1988), Paul Gilroy (1993), Dipesh Chakrabarty (1992, 2002a, 2002b, 2008 [2000]), Valentin Mudimbe (1988), Mahmood Mamdani (1996), Michel-Rolph Trouillot (1991, 2002, 2003), Arturo Escobar (1994), Timothy Mitchell (2000) and Shalini Randeria (1999a, 1999b, 2002). These authors have been concerned with how to move beyond a relativist approach to the foundational binary constructions that have been characteristic of the discourse on modernity, such as modern/traditional, Western/non-Western, centre/periphery, civilized/primitive, or rational/irrational.

Shalini Randeria has proposed two concepts to develop alternative historical perspectives to the classic conception of modernity: *entangled histories* and *uneven modernities* (Randeria, 1999). She suggests replacing 'the idea of a homogeneous Western modernity travelling, for the most part imperfectly, to the rest of the world' with a 'more messy and complex picture' (Randeria, 1999a, 1999b). This would address some of the problems that lie concealed in the binary concepts that have sustained the idea of modernity.

> I would suggest replacing a 'history of absences' (Mamdani 1996), as in discourse of modernisation theory, or a history by analogy, as in discourses of alternative modernities, by a relational perspective [on modernity] which foregrounds processes of interaction and intermixture in the entangled histories of uneven modernities (Randeria, 1999a: 1999b) (in Conrad & Randeria, 2002: 287).

Randeria's notion of 'entangled histories' makes a case for substituting the binary categories of modernity with a new analytical framework in which the units of analysis emphasize the constitutive role of 'the exchange and flow of ideas, institutions and practices' (Conrad & Randeria, 2002: 8). Her second concept, 'uneven modernities' specifies the focus of these new units of analysis. Rather than identifying historical commonalities in this exchange and interaction, the framework aims to reveal 'demarcations and fractures' in the texture and constitution of the modern world (Conrad & Randeria, 2002: 18/9).

In STS, both the singular concept of modernity (e.g. Chambers & Gillespie, 2000; de Laet & Mol, 2000; Campion & Schrum, 2004; Latour, 1993, 2007; Hess, 2001, 2007; Thompson, 2008) and the relativist approach of multiple, plural or alternative modernities (e.g. Adams, 2002; Redfield, 2002; Anderson, 2002; Harding, 2008;[254] Anderson & Adams, 2008) are prevalent. For example, in 2008, Anderson and Adams lamented that various attempts to pluralize the concept of modernity in the 1990s have been 'largely ignored' by scholars in science and technology studies (Anderson & Adams, 2008: 183). Along with other STS scholars,[255] they view the field of 'post-colonial technoscience' as a promising site for dealing with 'the post-colonial critique

---

254 Sandra Harding has recently drawn attention to the 'under-addressed' modernity/ tradition binary in Science and Technology Studies (Harding, 2008: 7).

255 Special Issue on Postcolonial Technoscience in *Social Studies of Science*, Vol. 32, No. 5/6 (Oct–Dec, 2002), Special Issue on Postcolonial Technoscience in *Science as Culture*, Vol. 14, No. 2, 2005.

that informs the anthropology of modernity' (Anderson & Adams, 2008: 183). Its objective, in their view, is to follow the movement of science and technology into new environments (Anderson & Adams, 2008: 183) by pursuing 'multi-sited histories of science' (Anderson & Adams, 2008: 192).[256]

This focus on the traffic or travel of knowledge is an example of the kind of perspective that Mitchell (2000), Randeria (2002), Bhambra (2011) and others see as a historiographical frame or grand narrative of world history – a perspective in which the single conceptualization of modernity ironically remains intact. The standard analytical procedure from this perspective, according to Randeria, is doomed to compare the Western with (the deficient) non-Western historical experience (Randeria, 2002).[257] Chakrabarty refers to the kind of 'historical construction of temporality' which separates the modern from the premodern by historical time (Chakrabarty, 1992: 13) as a mode of writing history as a transition narrative. The transition narrative inscribes a particular 'structure of global historical time' that situates the modern individual at the very end of history. The mode of writing history as 'first in Europe, then elsewhere' (Chakrabarty, 2008 [2000]: 10), in Chakrabarty's theory, represents a rehearsal of the split between the modern and the pre- or non-modern: '[...] this split is what is history;

---

[256] 'We need multi-sited histories of science which study the bounding of sites of knowledge production, the creation of value within such boundaries, the relations with other local social circumstances, and the traffic of objects and careers between these sites, and in and out of them. Such histories would help us to comprehend situatedness and mobility of scientists, and to recognize the unstable economy of "scientific" transaction. If we are especially fortunate, these histories will creatively complicate conventional distinctions between center and periphery, modern and traditional, dominant and subordinate, civilized and primitive, global and local' (Anderson & Adams, 2008: 192).

[257] 'The usual mode of engaging in a comparative exercise idealises and abstracts from Western experience in order to then compare (more often than not negatively) non-Western trajectories, transformations and institutions of civil society as deficient or different. These narratives, whether Marxist or liberal, view social reality through the lens of binary oppositions (West/non-West, modern/traditional, societies with history/societies without history, secular/religious). Non-Western societies, as the very term signifies, are defined by negation. [...] Their historical and contemporary experience is then understood in such a framework not in terms of what it is but in terms of what it is not' (Randeria, 2002: 291).

writing history is performing this split over and over again' (Chakrabarty, 1992: 13). The transition narrative appeared in the study of Hughes' theory of technological change and Shapin and Schaffer's study of experimental life, which demonstrated that modern science and technology require an alterity against which they take their full meaning.

5.3 The Programme in Science Studies Elsewhere

Trouillot's notion of *Elsewhere* and Randeria's concepts of *entangled histories* and *uneven modernities* (Randeria, 1999a: 1999b) can be used to study the constitutive role of alterities in modern scientific and technological culture. The notion of Science & Technology Studies Elsewhere is coined to connect these concepts from postcolonial studies, anthropology and cultural studies for this purpose. This is not simply an isolated intellectual exercise. On the contrary, it follows similar attempts to come to terms with the meaning of recent debates on modernity in other disciplines and fields in the social sciences and humanities, such as Costa (2005) and Boatcă et al. (2010) in sociology, and Trouillot (1991, 2002) and Restrepo and Escobar (2005) in anthropology.

Hughes' *Networks of Power* and Shapin and Schaffer's *Leviathan and the Air-Pump* implement principles of a research programme in the Sociology of Scientific Knowledge, associated with an Edinburgh school of STS (for example, Barnes, 1977; Bloor, [1976] 1991; Barnes and Bloor, 1982). One of the founding texts of the Strong Programme is David Bloor's book *Knowledge and Social Imagery* (Bloor, [1976] 1991). This book, too, has been influential in STS; Martin et al. list it as number nine in core STS literature (Martin et al., 2012). In this book, Bloor formulated four tenets for the Strong Programme; causality, impartiality, symmetry and reflexivity. The implications of the postcolonial examination of Hughes' and Shapin and Schaffer's books for the field of STS can be thought through by clarifying their relation to the Strong Programme. These consequences are articulated by formulating four tenets in the Programme in Science Studies Elsewhere in relation to Bloor's version of the Strong Programme (Bloor, [1976] 1991).

The Programme in Science Studies Elsewhere casts science and technology as *North Atlantic universals* that can be studied through Trouillot's intertwined lenses of the geography of management and imagination. Elsewhere makes visible the entangled processes and practices that inscribe sameness and difference in modern science and technology. The Programme intends to draw attention to the default replication of this production of alterity by STS. It is not proposed as a call to equalize the global political economy of STS (this important case has long been made, though undoubtedly it would need to be restated). The table below presents the four tenets of the Programme in Science Studies Elsewhere in relation to Bloor's four tenets for the Strong Programme in SSK (Table 7).

The first tenet: Shapin and Schaffer's first tenet for the Strong Programme, causality, leads them to inquire into the historical circumstances in which modern experimental life became institutionalized. Their aim is to make visible a direct *causal* connection between Restoration society and the rules, procedures and conventions that define experimental practice in the laboratory.

The study of Elsewhere in Hobbes and Boyle's philosophical programme shifts the analytical focus away from causal linkages in the study of experimental culture to the entangled processes that produce scientific meaning and knowledge against a contrastive alterity. The analysis inquired after the Elsewhere of Boyle and Hobbes's philosophical programmes, the oppositional referents outside (and within) themselves, the pre- or non-modern science in relation to which their philosophical programmes attain their full meaning. The complementary Elsewhere to the Here of their philosophical programmes was the New World, the necessary alterity for the construction of the beginnings of a common European self, both of which premised one another and were conceived as inseparable. 'Inferior creatures' and 'savages' were the relational Others that inhabited the Elsewhere in their programmes. Elsewhere can serve as an analytical tool to emphasize the *entangled* processes and practices of othering that have shaped modern science.

The second tenet: Shapin and Schaffer's approach to studying experiment aims at delivering an *impartial* account of rejected (Hobbes) and accepted (Boyle) knowledge. Their distinction between member's and stranger's accounts allows them to reveal causal links between Hobbes and Boyle's respective knowledge claims and historical judgements in seventeenth century England. In their historical approach, the imaginary figure of the 'genuine', 'ignorant stranger' is assigned a permanent place to occupy outside of

The Strong Programme in the Sociology of Scientific Knowledge (Bloor 1991 [1976])	The Programme in Science Studies Elsewhere (Hofmänner, 2016)
(1) It would be *causal*, that is, concerned with the conditions which bring about belief or states of knowledge. Naturally, there will be other types of causes apart from social ones which will co-operate in bringing about belief.	It would be *entangled*, that is, concerned with the constitutive role of Elsewhere in the practices and processes that shape knowledge claims in modern science.
(2) It would be *impartial* with respect to truth and falsity, rationality or irrationality, success and failure. Both sides of these dichotomies will require explanation.	It would be *relational* with respect to claims for chronological primacy and universality in modern science. The shaping of binary categories of historical thinking about modern science (modern/traditional, rational/irrational, knowledge/non-knowledge, etc.) will require relational (rather than oppositional) explanation.
(3) It would be *symmetrical* in its style of explanation. The same types of cause would explain, say, true and false beliefs.	It would be *uneven* in its style of explanation. Its units of analysis would foreground demarcations and fractures in the social imaginary of modernity to explain, say, true and false beliefs.
(4) It would be *reflexive*. In principle, its patterns of explanation would have to be applicable to sociology itself. Like the requirement of symmetry, this is a response to the need to seek for general explanations.	It would be *contestable*. In principle, its patterns of explanation would have to allow critique that contests knowledge claims on the Here and the Elsewhere. This is a response to the need to develop and debate entangled modes of explanation.

Table 7: The four tenets of the Programme in Science Studies Elsewhere in reference to Bloor's four tenets of the Strong Programme in the Sociology of Scientific Knowledge (Hofmänner, 2016).

experimental culture from the time of its inception in early Europe all the way through to the present. He/she fills the necessary slot, personifies the indispensable (historical) absence against which the presence of Shapin and Schaffer's knowledge claims can be contrasted and instantiated.

Shapin and Schaffer's inscription of the oppositional members and strangers to experimental culture is a rehearsal of Chakrabarty's 'transition narrative' – an inscription of historical time as a measure of cultural difference (Chakrabarty, 2008 [2000]). This rehearsal feeds on replicating Elsewhere. Shapin and Schaffer's 'ignorant stranger' is assigned to Chakrabarty's *Imaginary Waiting Room* of history (Chakrabarty, 2008 [2000]). He/she is waiting there for modern science and its experimental culture to travel to him/her so that he/she can acquire membership in the universal experimental culture. However, he/she can never become a member of this heritage because he/she is doomed to represent the contrastive outside against which the inside identity of experimental culture is constituted. Shapin and Schaffer's impartial approach of 'playing' the 'ignorant stranger' may be seen as a rhetorical device for performing Chakrabarty's historical split, the historical construction of a temporal axis that constantly reconstitutes the outside (Chakrabarty, 2008 [2000]) of modern science.

The Programme in Science Studies Elsewhere claims that the dichotomies of Bloor's Strong Programme (truth and falsity, rationality or irrationality, success and failure) form part of a historiography of science and knowledge that depends upon the twin concepts of alterity and identity. The relational approach to modern science seeks explanations for the processes and practices that inscribe such opposing categories of historical thinking. Emphasizing the role of Elsewhere in the making of modern science is an attempt to work against Chakrabarty's perpetual historical split by shifting attention to the relational entanglement of the dichotomies of modernity that have favoured certain knowledge claims over others.

The third tenet: Shapin and Schaffer apply a *symmetrical* style of explanation to the knowledge claims of Boyle and Hobbes to deliver an impartial account of the two sides of the dichotomies truth/falsity, rationality/irrationality, success/failure. The analysis of Elsewhere in Boyle and Hobbes's philosophical programmes has shown that solving the problems of knowledge and order necessitated a contrastive background of non-knowledge and disorderliness (the New World, its 'inferior creatures' and 'savages') to legitimize knowledge claims. But this contrastive background of alterity does not simply introduce another analytical level in need of symmetrical inspection. A symmetrical explanation of the constrasting knowledge claims of Hobbes/ Boyle and 'inferior creatures'/ 'savages' is bound to generate a 'history of absences' (Mamdani 1996). Symmetrical postulates on the foundational, binary oppositions of modernity re-inscribe how difference has been codified in modern science.

The Programme in Science Studies Elsewhere proposes to replace Bloor's symmetrical style of explanation with an *uneven* style of explanation. It attempts to shift the focus away from approximating an ideal image of the symmetrical neutral view from nowhere to deconstruct modern science. It does not attempt to eliminate skewed power relations of divided or shared histories. Neither does it simply try to establish historical commonalities between Boyle and Hobbes's philosophical programmes, experimental practice, the Royal Society, travellers, tradesmen and colonial officers to the New World, the English East India Company, the Christian Virtuoso, 'savages', and 'inferior creatures'. The uneven style of explanation seeks new units of analysis that provide transversal accounts of knowledge claims to investigate the practices and processes that underlie the binary categories of modernity.

Following Randeria (2002), the shift in style of explanation involves a change of perspective on the status of modern science: instead of representing a historical-philosophical category of modernity, modern science becomes an agent in staging the social imaginary of modernity. The Programme in Science Studies Elsewhere seeks to reveal the messy demarcations and fractures in the texture and constitution (Conrad & Randeria, 2002) of this agency. Studying the unevenness of the foundational categories of modern science opens the scene for alternative social imaginaries of knowledge.

The fourth tenet: Bloor's last tenet on *reflexivity* specifies that Shapin and Schaffer's patterns of historical explanation of experimental culture would have to be applicable to sociology itself. Shapin and Schaffer do not address this tenet in their study. However, from the point of view of Science Studies Elsewhere, Bloor's fourth tenet illustrates the vicious circle of explanation within which STS has been operating. At the outset of this book, calls were quoted to address STS's paradoxical condition of representing itself by means of an intellectual genealogy that remains detached from its 'broader historical and socio-political contextuality' (Elzinga, 1997) while at the same time arguing for the co-construction between research and subject for other knowledge fields (Guggenheim & Nowotny, 2003). Could not a reflexive perspective provide a solution to this paradoxical condition? Would not the application of STS's standards and explanations to its own programmes and concepts provide an analytical framework for examining its own co-construction?

A reflexive perspective is bound to reproduce STS's knowledge claims on the Here and the Elsewhere. This perspective would have to remain caged in this field's foundational assumptions on modernity. New approaches are necessary to challenge these assumptions. In this examination, Elsewhere (and its associated analytical lenses of the geographies of management and imagination) was proposed as an analytical tool for demonstrating an alternative way of framing the phenomenon of experimental culture and practice. The Programme in Science Studies Elsewhere proposes that debates on alternative ways of framing science be a constitutive requirement for STS in the future. Suggesting that the Programme be *contestable* is to accentuate the added value of engaging with such alternative ways of mapping the Here and the Elsewhere of knowledge claims.

Trouillot's notion of Elsewhere may be used as a conceptual tool to call attention to characteristics and procedures of science and technology that have remained systematically eclipsed. Science & Technology Studies Elsewhere delineates a terrain for inquiry into knowledge claims that views the construction of alterity and identity as key strategies of modern science. To study these strategies, a Programme in Science Studies Elsewhere is proposed in reference to David Bloor's Strong Programme in the Sociology of Scientific Knowledge, which provided the theoretical backbone to Shapin and Schaffer's methodological approach to experiment. Where Bloor proposes a causal, impartial, symmetrical and reflexive Strong Programme, this study proposes an entangled, relational, uneven and contestable Programme.

The above analysis might be dismissed as providing material that is relevant only to specialist fields like the anthropology of science, cultural studies of science, or postcolonial technoscience. However, Science & Technology Studies Elsewhere attempts to move beyond the project outlined by postcolonial technoscience, the pluralizing rhetoric of which ironically cements the universal historical narrative of a uniform modernity and its European heritage. Its primary object is not to analyse modern science through the lens of the cultural or historical diversity ascribed to plural or alternative modern-ities. It does not propose to develop strategies for overcoming or doing away with Elsewhere as a device for the construction of meaning in scientific prac-tice. Nor does it challenge its function in the scientific quest for cognitive hegemony – such approaches would not address the basic problems and assumptions that underlie the key explanatory categories of modernity (modern/traditional knowledge, public/private sphere, etc.) (Bhambra, 2011). The challenge is not to recognize the difference of Shapin and Schaff-er's genuine, ignorant stranger but to question his unrivalled status and func-tion as an oppositional referent to sustain the imaginary geographies of iden-tity and coherence in modern science.

Viewed through Trouillot's lens on Elsewhere, the field of STS does not much resemble the heterogeneous, multi-sited, diverse interdisciplinary field it has been promoted as. Rather it appears as a homogeneous, concerted re-inscription of the assumptions and binary categories that underlie the grand narrative of modernity. As in other fields in the social sciences and human-ities, STS has built its academic identity by casting its object of study in rela-tion to the concept of modernity, a now contested concept. Its contested sta-tus requires fresh perspectives that are able to ask new questions. The Programme in Science Studies Elsewhere posits the need for a cognitive shift in perspective on modernity as the imaginary frame against which STS has legitimized its subject. It seeks to move the analytical focus of STS to study-ing and contesting the strategies that have successfully established claims on modern science and knowledge.

But why should STS perform this move? On what grounds would it be sensible to compromise STS's present course, given its relative success in the past? Why should a seemingly abstract expression such as Science & Tech-nology Studies Elsewhere be taken seriously? This chapter argues in favour of undertaking research on Science & Technology Studies Elsewhere as an exercise in addressing Elzinga's 'paradox' (1997) and Guggenheim and No-wotny's (2003) 'collective Blind Spot' of STS, respectively. Anderson's (2007) vision of re-charting the map of STS for the twenty-first century cannot commence as long as STS eclipses its collective Blind Spot.

However, a plain collage of STS's genealogy against its specific Western European and North American social, political and cultural context will not assist in better understanding STS's specific conditioning nor will it do away with the collective Blind Spot. The formative context for STS was given not only by the sum of the internal social, political and cultural specificities of North America and Western Europe, but also by the conceptual confines of the assumptions that underlie the grand narrative of modernity. These assumptions rely on imaginary geographies that produce a space for the Other, Elsewhere. The travels of STS to other countries, as envisioned by Anderson in the Introduction, appear as a rehearsal of Chakrabarty's 'first in Europe, then elsewhere' (Chakrabarty, 2008 [2000]: 7) structure of global historical science. This transition narrative recreates the binary categories of thinking about the modern world: modern/traditional, public/private and Western/non-Western. It situates the North American and Western European STS scholar at the very end of history and consigns the (imagined) Rest to Chakrabarty's *Imaginary Waiting Room* of history (Chakrabarty, 2008 [2000]).

Corbey and Leerssen claimed that there are various ways in which Otherness can be portrayed and that the use of cultural alterity '[...] does not by definition imply a denigration of the Other' (Corbey & Leerssen, 1991b: vii). Science & Technology Studies Elsewhere provides an analytical terrain for seeking such alternative ways of casting knowledge claims. Perhaps the Elsewhere of modern science is more easily discerned by the inhabitants of the (at times imaginary) spaces it has produced, such as that of Shapin and Schaffer's genuine, ignorant stranger to experimental culture.

6. The Revolving Door of STS

The Programme in Science Studies Elsewhere was developed with reference to two specific books that were written in the mid-1980s, and its central tenets were articulated in relation to a particular intellectual tradition that shaped STS at the time. It remains to be shown that the Programme's tenets are relevant to subsequent STS scholarship. For this purpose, this chapter consults a third well-known STS book, Bruno Latour's celebrated treatise *We Have Never Been Modern* (1993). This book qualifies as a suitable candidate because it refers to both Shapin and Schaffer's *Leviathan and the Air-Pump* and Hughes' *Networks of Power*. Latour is viewed by many as having been 'the dominant influence within the field of STS' (Martin et al., 2012: 10). His work, together with that of Michel Callon and John Law, among others, has contributed to the development of Actor Network Theory (ANT), a strand of STS that has been extremely successful. When STS travels to global places outside of the North Atlantic, Latour's book will most likely be part of its baggage.

In *We Have Never Been Modern*, Latour presents his model of modernity. His point of departure for this project is Shapin and Schaffer's *Leviathan and the Air-Pump*. Latour also refers to Hughes' study on electrification and uses the metaphor of the seamless web and the notion of the network. For this reason, Latour's book provides an opportunity to trace the findings of our two paradigmatic STS case studies in subsequent STS scholarship.

The chapter will show that Latour follows in Shapin and Schaffer's analytical footsteps in playing the ignorant stranger to modern culture. Therefore, just as Shapin and Schaffer's stranger to experimental culture replicated the alterities of modern science, Latour's model of modernity evokes an Elsewhere, a space for the Other – despite his assertions to the contrary. Furthermore, the discussion will demonstrate that Latour deployed a modified version of Hughes' metaphor of the seamless web to erase conceptual boundaries between nature and culture, and between humans and nonhumans. However, even though he ascribes to the seamless web several orderly and appeasing properties, his book retains the assumptions that underlie the original metaphor. In this way, just like Hughes, Latour engenders a Savage Slot and configures (imaginary) people outside his scope of analysis.

Latour's example demonstrates that Elsewhere and Seamlessness can be traced in STS scholarship subsequent to Hughes', and Shapin and Schaffer's books. To consider the implications of their persistence in STS's body of knowledge the chapter reflects on their effect on the STS novice who is keen to become part of the STS community.

6.1 Latour's Natureculture

Latour's Elsewhere: Playing the Stranger

Shapin and Schaffer's *Leviathan and the Air-Pump* is the centrepiece of Latour's *We Have Never Been Modern* (1993). Latour builds his argument, hypotheses, and conclusions upon Shapin and Schaffer's study and the classic debate between Robert Boyle and Thomas Hobbes. In his view, Boyle and Hobbes were 'arguing over the distribution of scientific and political power' and in the process invented our modern world (Latour, 1993).[258] The modern world is defined by the modern Constitution (with a capital C). Latour considers the debate between 'the natural philosopher' Robert Boyle and 'the political philosopher' Thomas Hobbes in the middle of the seventeenth century as 'an exemplary situation that arose at the very beginning' of the drafting of this modern Constitution (Latour, 993: 15). This backdrop sets the scene for Latour's key argument: that the modern Constitution has to be revised.

To solve the modern dilemma that followed from Boyle and Hobbes's invention, Latour intends to define a 'new anthropology' (Latour, 1993: 17) and a new, 'complete Constitution' (Latour, 1993: 29). The new

---

[258] In this world, 'the representation of things through the intermediary of the laboratory is forever dissociated from the representation of citizens through the intermediary of the social contract' (Latour, 1993: 27).

anthropology can be developed by using the 'ideal laboratory material' provided by the 'disagreements' between Boyle and Hobbes (Latour, 1993: 17). In turn, the revised Constitution can be accomplished by '[generalizing] the results achieved by Shapin and Schaffer' (Latour, 1993: 30).

Latour begins his book by citing instances from an imaginary newspaper, in which he identifies elements of the 'modern paradox'. The modern paradox has to do with the division between nature and culture, an arrangement that is subjected to constant efforts to keep these spheres apart and to tie them together.[259] Furthermore, the modern Constitution 'believes in the total separation of humans and nonhumans' but 'simultaneously cancels out this separation'. These opposing properties 'made the moderns invincible' (Latour, 1993: 28). Latour introduces terms such as 'purification' and 'translation', 'networks', 'hybrids' and 'quasi-objects' to capture the dynamics that lie beyond this artificial division. Finally, based on his discussion of Shapin and Shaffer's book, Latour proposes a 'Parliament of Things' as a solution to the division of powers that was imposed by the modern Constitution. Latour's expressions have been exceptionally successful in the field of STS and continue to be widely applied in STS research and teaching.

Shapin and Schaffer's celebrated conclusion of *Leviathan and the Air-Pump* – that 'solutions to the problem of knowledge are solutions to the problem of social order' (Shapin & Schaffer, 2011 [1985]: 332) – provided Latour with an excellent starting point for devising his model of modernity. This conclusion had established in the STS community the power of the

---

259 'Here lies the entire modern paradox. If we consider hybrids, we are dealing only with mixtures of nature and culture; if we consider the work of purification, we confront a total separation between nature and culture. It is the relation between these two tasks that I am seeking to understand. While both Boyle and Hobbes are meddling in politics and religion and technology and morality and science and law, they are also dividing up the tasks to the extent that the one restricts himself to the science of things and the other to the politics of men. What is the intimate relation between their two movements? Is purification necessary to allow for proliferation? Must there be hundreds of hybrids in order for a simply human politics and simply natural things to exist? Is an absolute distinction required between the two movements in order for both to remain effective? How can the power of this arrangement be explained? What, then, is the secret of the modern world?' (Latour, 1993: 29).

symmetry principle of the sociology of scientific knowledge (SSK) for the historical passage to modernity. From this vantage point, Latour was able to put forward the assertion that the two equal opponents, Boyle and Hobbes, invented the first 'Great Divide' between things and humans and the second 'Great Divide' between moderns and premoderns.[260] They also invented the modern Constitution that separates scientific and political power to represent things and subjects, respectively (Latour, 1993: 28). Latour sets out to deconstruct these two Great Divides.

However, Latour's use of Shapin and Schaffer's study alone does not suffice to demonstrate that he continues their replication of the alterities of modern science. After all, Latour criticizes their book at great length. Furthermore, Latour advocates symmetry: symmetry between nature and culture, symmetry with respect to 'Westerners' and 'Others', symmetry between 'Us' and 'Them', symmetry in anthropology 'at home' and in 'the tropics'. Indeed, Latour sets out to erase precisely these boundaries and tells us that, in reality, there is no distinction between nature and culture, between Us and Them, no 'Westerners' and 'Others', there is no modern or premodern. Therefore, how is it possible for Latour's *We Have Never Been Modern* to replicate the alterities of modern science that were traced in Shapin and Schaffer's *Leviathan and the Air-Pump?*

The answer may be found in Latour's methodology. Latour dismisses the categories of the moderns and the premoderns and invokes a third position that allows him to contemplate both of these categories. This view from the outside corresponds to Shapin and Schaffer's method of playing the stranger to modern culture. This approach assumes that 'playing' the stranger configures an outside view from nowhere. Yet the actor of Latour's third position is, in reality, modern, and not premodern. He or she retains their cultural heritage with Boyle and Hobbes. Latour's use of personal pronouns reveals this predicament.

Latour sometimes speaks about 'our modern world' in contrast to the premodern world of the others. At other times, Latour dissociates his own

---

260   'The internal partition between humans and nonhumans defines a second partition – an external one this time – through which the moderns have set themselves apart from the premoderns' (Latour, 1993: 99).

person from the moderns. For example, he laments that 'the price the moderns paid for this freedom [...] was that they remained unable to conceptualize themselves in continuity with the premoderns. They had to think of themselves as absolutely different, they had to invent the Great Divide' (Latour, 1993: 39). Latour's position is variable: sometimes he belongs to the moderns, sometimes he does not; but this position is also selective. On no occasion does he position himself with the premoderns. He always refers to 'other nature cultures' (Latour, 1993: 11) or 'other culture-natures' (Latour, 1993: 38) as premoderns and always uses the personal pronoun 'them'. Latour concludes that 'we are no longer entirely modern' but that 'we are not premodern either' (Latour, 1993: 127).

Nevertheless, Latour's solution is not to return to the 'natureculture' state of the premoderns: 'we do not wish to become premoderns all over again' (Latour, 1993: 140). Latour's solution to the modern paradox is to amend the Constitution and to constitute a Parliament of Things. However, interestingly, this solution does not abandon the modern but rather calls for a ratification of 'what we have always done':

> However, we do not have to create this Parliament out of whole cloth, by calling for yet another revolution. We simply have to ratify what we have always done, provided that we reconsider our past, provided that we understand retrospectively to what extent we have never been modern, and provided that we rejoin the two halves of the symbol broken by Hobbes and Boyle as a sign of recognition. Half of our politics is constructed in science and technology. The other half of Nature is constructed in societies. Let us patch the two back together, and the political task can begin again (Latour, 1993: 144).

In other words, the new Constitution and the Parliament of Things require that the model dividing nature from culture, the modern from the premodern, as well as the new entities Latour has placed into these divides, be retained. The solution is to come to terms with these new concepts, to change their focus and dynamics. Once again, the question is, who is Latour speaking about when he uses the personal pronoun 'we'? Who will 'ratify what we have always done'? The following section offers some indications about these agents.

> In this desire to bring to light, to incorporate into language, to make public, *we continue to identify with the intuition of the Enlightenment*. But this intuition has never had the anthropology it deserved. It has divided up the human and the non-human and believed that the others, rendered premoderns by contrast, were not supposed to do the same thing. [...] We have been modern. Very well. We can no longer be modern in the same way. When we amend the Constitution [...] (Latour, 1993: 142, my emphasis).

After he has eliminated the categories of the modern and the premodern, Latour bizarrely returns to retain a worthy kernel of the moderns. He separates out a particular element of modern identity, 'the intuition of the Enlightenment', which was in principle correct, but unfortunately was served by a faulty anthropology. The crucial issue in Latour's confusing use of 'us' and 'them' concerns to whom Latour assigns agency for change. Who will amend the Constitution and constitute the Parliament of Things? Who continues 'to identify with the intuition of the Enlightenment'?

Latour assigns agency for change to those who 'have been modern'. They are his audience. Latour does not speak to the Others, the ones who have not inherited the repertoire of Hobbes and Boyle's legacy, as revealed in the following quote:

> The task of our predecessors was no less daunting when they invented rights to give to citizens or the integration of workers into the fabric of our societies. [...] It is up to us to change our ways of changing (Latour, 1993: 145).

Latour's and the readers' predecessors are Hobbes and Boyle; their predecessors are heirs to the separation of nature and culture. By this very distinction, Latour's audiences are successors to the moderns. His third position retains a common heritage with Hobbes and Boyle. Just like Shapin and Schaffer, he views the seventeenth-century ideas of Boyle and Hobbes as the beginnings of the cultural heritage of modern science. He depicts Hobbes ('and his disciples') and Boyle ('and his successors') as inventors of the 'major repertoires' of words used by the moderns (Latour, 1993: 25). These repertoires include words such as 'representation, sovereign, contract, property, citizens' (Hobbes) and 'experiment, fact, evidence, colleagues' (Boyle) (Latour, 1993: 24–5). Moreover, Latour views Boyle and Hobbes as resembling 'a pair of Founding Fathers' to the new Constitution (Latour, 1993: 28) which 'invents a separation between the scientific power charged with representing things and the political power charged with representing subjects' (Latour, 1993: 29).

Latour's third position only becomes effective by contrasting his cultural heritage against others, who do not form part of the cultural heritage of Hobbes and Boyle. As the study by Shapin and Schaffer has shown, this position does not do away with the construction of alterities, on the contrary. The acting stranger also configures the category of the 'ignorant stranger'. The true stranger is ignorant, while the acted stranger is knowledgeable by virtue of his cultural association with the moderns; he is only 'playing' the stranger.

No matter how elusive Latour's position appears with regard to the 'moderns' and the 'premoderns', his analytical stance abides by the cultural heritage of Boyle and Hobbes. Even if Latour asserts that categories such as moderns and the nonmoderns, and 'Westerners' and 'Others' have become redundant, his own affiliation with the cultural heritage of Hobbes and Boyle invokes an alterity. Latour's text advocates a new 'symmetrical anthropology' to overcome the opposing categories of 'Westerners' and 'Others' (Latour, 1993: 104). However, his text in fact re-inscribes these categories, replicating the alterities that have constituted modern science, despite his verbatim assurances to the contrary. Latour assumes a symmetrical style of explanation to approximate a neutral analytical position by playing the stranger to modern culture. This ideal neutral view from nowhere to deconstruct science re-inscribes the binary categories that it sets out to eliminate.

Furthermore, Latour adheres to the transition narrative that structures historical time as emanating from Europe. He follows a specific chronology of events to recount the genesis of the two Great Divides. The moderns were created by the first, the Internal Great Divide between nature and culture. In turn, this divide produced the second, the External Great Divide between the moderns and the premoderns.[261] The moderns came first. This chronology is

---

[261] 'So the Internal Great Divide accounts for the External Great Divide' (Latour, 1993: 99) and '[i]n order to understand the Great Divide between Us and Them, we have to go back to that other Great Divide between humans and nonhumans that I defined above' (Latour, 1993: 97).

faulty; premoderns were not invented after the moderns.[262] The two ideas presupposed each other and therefore were conceived in tandem. Just like Shapin and Schaffer, Latour overlooks the point that the 'premodern' was not a consequence of modernity, but that the construction of alterity was constitutive of the very idea of modernity. His transition narrative has consequences for the travels of STS.

Latour's Seamlessness: More of the 'Savage Slot'

In *We Have Never Been Modern*, Latour also applies concepts that Hughes had introduced in his systems model of technological change. In particular, Latour makes extensive use of the metaphor of the seamless web and the notion of the network. On occasion, Latour's text refers to Hughes' work on electrification but he does not acknowledge any intellectual inheritance from Hughes. Instead, Latour presents his own version of the seamless web and the network as components of his own model. Nevertheless, the connection between his book and Hughes' concepts is easily verified. Latour uses the word 'network' in conjunction with the metaphor of the seamless web. There is no dispute about the intellectual origins of this metaphor in STS. Hughes introduced the seamless web in conjunction with his systems or network approach. The association between the network and the seamless web establishes the intellectual lineage between Latour and Hughes. For this reason, the book provides an excellent example to probe manifestations of Seamlessness in STS.

Latour employs the metaphor of the seamless web to characterize 'the fabric' of the moderns (Latour, 1993: 7). The moderns have created the analytic distinction between nature and culture. Beyond this division, Latour

---

262   'Century after century, colonial empire after colonial empire, the poor premodern collectives were accused of making a horrible mishmash of things and humans, of objects and signs, while their accusers finally separated them totally – to remix them at once on a scale unknown until now [...]. As the moderns also extended this Great Divide in time after extending it in space, they felt themselves absolutely free to give up following the ridiculous constraints of their past which required them to take into account the delicate web of relations between things and people' (Latour, 1993: 39).

recognizes 'the seamless fabric' of what he calls 'nature-culture' (Latour, 1993: 7). Analytic continuity, in his view, is only possible when this fabric is seamless. To return to this ideal state of the seamless fabric of nature-culture, he proposes writing a new Constitution. He assigns this task to the anthropologists, since they are 'accustomed [...] to dealing calmly and straightforwardly' with this seamless fabric (Latour, 1993: 7). Unfortunately, however, the anthropologists are only able to deliver this analytic continuity when they '[go] off to the tropics to study others' abroad and not 'at home' (Latour, 1993: 7).[263]

As noted above, Latour does not declare his intellectual debt to Hughes but accords the network a different meaning and employs the seamless web metaphor for a different purpose. Why did Latour choose these notions only to change their original meaning and purpose in his model? What is their function in his text? The discussion below will show that Latour used these notions because they imply Seamlessness.

The metaphor of the seamless web structures Latour's imaginary geography of modernity. He views the modern condition as a configuration of concepts that has been imposed onto the seamless fabric of nature-culture. In this model, the seamless fabric of nature-culture is the tabula rasa, the default condition freed of linguistic concepts. The new Constitution is needed to navigate this tabula rasa in new ways that transgress the binary categories of modernity, such as the moderns and premoderns. Latour introduces a number of new concepts that aim at re-establishing analytic continuity. His notion of the network provides the topology for this navigation.

If the argument of this study is correct, we need to demonstrate that in Latour's book the utopian orderliness implied by the seamless web engenders the distinction between the Here and the Elsewhere. In other words, we have to show that Latour's use of the seamless web and the network engenders Trouillot's 'Savage Slot' and banishes this category from historical writing.

---

263 'Send her off to study the Arapesh or the Achuar, the Koreans or the Chinese, and you will get a single narrative that weaves together the way people regard the heavens and their ancestors, the way they build houses and the way they grow yams or manioc or rice, the way they construct their government and their cosmology. In works produced by anthropologists abroad, you will not find a single trait that is not simultaneously real, social and narrated' (Latour, 1993: 7).

Latour's semantics are notoriously elusive and difficult to pin down. He does not define the network. He hints at the meaning of this word by using it in relation to other elements or properties of his model. However, these references are also vague, as the examples below illustrate. For instance, Latour sees the modern world as constituted by networks[264] but at the same time describes these networks as 'more invisible than spiderwebs' (Latour, 1993:4). Furthermore, beyond the network, there is nothing.

> Between the lines of the network there is, strictly speaking, nothing at all: no train, no telephone, no intake pipe, no television set. Technological networks, as the name indicates, are nets thrown over spaces, and they retain only a few scattered elements of those spaces. They are connected lines, not surfaces. They are by no means comprehensive, global or systematic, even though they embrace surfaces without covering them, and extend a very long way (Latour, 1993: 118).

Nonetheless, networks constitute the most important components of Latour's model, since they 'weave our world' (Latour, 1993: 8), and are 'full of Being' (Latour 1993: 66).[265] Networks provide explanations ('What reason complicates, networks explicate' (Latour, 1993: 104)), and guide us through the labyrinths of modernity. Latour conjures Greek mythology to emphasize the significance of networks.

> More supple than the notion of system, more historical than the notion of structure, more empirical than the notion of complexity, the idea of network is the Ariadne's thread of these interwoven stories (Latour, 1993: 3).

For Latour, Ariadne's thread corresponds to the thread of his networks. This thread weaves the seamless web. It has the power to confer continuity between the local and the global, between the human and the nonhuman.

---

264 'Seen as networks, however, the modern world, like revolutions, permits scarcely anything more than small extensions of practices, slight accelerations in the circulation of knowledge, a tiny extension of societies, minuscule increases in the number of actors, small modifications of old beliefs' (Latour, 1993: 48).

265 'Look around you: scientific objects are circulating simultaneously as subjects, objects and discourse. Networks are full of Being' (Latour, 1993: 66).

Yet there is an Ariadne's thread that would allow us to pass with continuity from the local to the global, from the human to the nonhuman. It is the thread of networks of practices and instruments, of documents and translations (Latour, 1993: 121).

Ariadne's thread also carries symbolic power. Not only does it confer analytical continuity, it also has the ability to solve problems; it guides the way out of the labyrinth. With its assistance, Latour proposes his solution to the question of the 'universality of science', one of the central challenges to relativist analyses of science and technology: How does science reach 'everywhere'? How does it become universal? (Latour, 1993: 24). Once more, Latour's solution is formulated in terms of networks: science reaches everywhere because 'the network of science is extended and stabilized' (Latour, 1993: 24).[266] Latour also uses the network to frame his argument for the need for a symmetrical anthropology: 'comparative anthropology becomes possible when networks receive 'a place of their own' (Latour, 1993:10). Finally, networks also figure in Latour's proposal to revise the modern Constitution.

Networks are ubiquitous in Latour's book. What purpose do they serve? How is it possible for Latour to develop a model around a notion so vaguely defined? Why does he use this term in conjunction with the seamless web? Latour strategically applies the network in conjunction with the seamless web to invoke the power of Seamlessness, which results from the elimination of analytical categories. It has a positive connotation, because it stages the removal of categories as a progressive intellectual act. Seamlessness projects clarity and orderliness; it implies the absence of errors and interruptions. Seamlessness assures continuity in time and space; it insinuates a condition of perfection. Simultaneously, the seamless web depicts a new kind of fabric, a conceptual tabula rasa. Conversely, its archetypal fabric configures Latour's analytical position outside of the framework of conventional categories. However, this archetypal state can only be discerned by the observer who has

---

[266] 'No science can exit from the network of its practice. The weight of air is indeed always a universal, but a universal in a network. Owing to the extension of this network, competences and equipment can become sufficiently routine for production of the vacuum to become as invisible as the air we breathe; but universal in the old sense? Never.' (Latour, 1993: 24).

transgressed the dichotomies of modernity by playing stranger to its cultural heritage. In other words, Latour uses the metaphor of the seamless web and the notion of the network as analytical strategy. However, the metaphor of the utopian seamless web engenders the Savage Slot. In turn, this residual category re-inscribes the very alterities Latour claims to have erased.

Latour's strategy has proven to be extraordinarily successful in STS. His book is widely cited in the STS community and his concepts and model are employed in empirical research. They have become an integral part of the intellectual identity of STS. When STS travels outside of the geographical region of the North Atlantic, these concepts and models form part of the intellectual baggage.

6.2 The Imaginary Waiting Room of STS

Latour's book *We Have Never Been Modern* illustrates that the notions of Elsewhere and Seamlessness can be traced in subsequent STS scholarship. They are carried along silently in STS's body of mainstream literature, concepts, and approaches. Of interest to this study, however, are the implications of Elsewhere and Seamlessness on the STS novice, who is learning the trade of STS somewhere outside Europe and North America.

The STS novice reading *We Have Never Been Modern* outside the North Atlantic is likely to be confused by Latour's use of personal pronouns. Which personal pronouns would this reader identify with? Latour's book is written for his community of STS scholars. By the mere act of reading Latour's book, the novice will be initiated into the customs and practices of STS. Latour's use of personal pronouns makes it clear that his book is addressed to the heirs of Boyle and Hobbes's cultural heritage; it is an exercise for them to play the stranger to the moderns. By implication, to become a member of the STS community, the novice will first have to appropriate this cultural heritage before being able to play the stranger. This silent rite of passage into STS is an implication of STS's Collective Blind Spot, which appears more pronounced when STS travels outside the North Atlantic, where Latour's personal pronouns are not self-evident.

The silent rite of passage leads the STS novice out of the *Imaginary Waiting Room* of history and into the STS community. This is possible because the Waiting Room is a conceptual space and not a geographical place. Its residents are imaginary constructs. The rite of passage into STS is determined by an intellectual topology and not by geographical regions. In other words, the residents of the *Imaginary Waiting Room* of history do not necessarily represent particular geographical regions, but form part of an imagined space. Scholars outside this geographical region of the North Atlantic are able to leave the *Imaginary Waiting Room* and enter the community of STS. Indeed, this study started by pointing out that the STS community outside of Europe and the North Atlantic has been growing. However, the rite of passage is more difficult to complete outside of the geographical regions of the North Atlantic, because novices are likely to be confused by Latour's use of personal pronouns.

To qualify as a member of the STS community, the novice will need to adopt the cultural heritage of Hobbes and Boyle. Latour's book will also teach the novice how to play the stranger to this culture. The novice will also be trained to apply the metaphor of the seamless web to instate clarity and orderliness by eliminating categories. Indeed, Latour's book reads as a map to navigate the rite of passage through the revolving door into STS. However, the STS novice is likely to be unaware that this particular revolving door leads through the passageway of the North Atlantic (Trouillot, 2003). Here, at this junction, the transition narrative of world history is conceived that situates the modern individual at the very end of history (Chakrabarty, 1992). The revolving door stages Europe as the single centre 'in reference to which all other regions are to be located; and an understanding of history in which there is only one unfolding of time, the history of the West, in reference to which all other histories must establish their significance and receive their meaning' (Mitchell, 2000: 7).

After having read the first few pages of *We Have Never Been Modern*, the STS novice is likely to wonder which newspaper Latour is quoting from: is it a French newspaper? What if the newspaper were written somewhere else, in a different language, for example, in Zulu or Telugu? What problems would Latour discern in these newspapers? How would they compare with the problems Latour quotes at the beginning of his book, the problems he has set out to solve?

These questions may be rephrased by returning to the mental picture at the beginning of this study. What would Latour see from the 'Top of Africa' skyscraper in Johannesburg, through his scholarly spectacles with lenses fitted for his study of modernity? This study suggests that a blind spot has settled on Latour's spectacles. Will he discern a seamless web in the City of Gold? Will he identify the inheritors of the cultural legacy of Hobbes and Boyle, heirs to these Founding Fathers? If not, what otherwise?

7. The March for Science, and STS

7.1 Times of crisis

The March for Science on 22 April 2017 has been portrayed as emblematic for the contemporary crisis in the relationship between science and society. Prominent historians of science have adjudicated the event as unprecedented. The participating scientists and protestors demonstrated to restore trust in science and expertise, to defend the freedom of science, and to demand that scientific facts be distinguished from untruths. Although these topics fall squarely within the scope of STS, this scholarly community has not taken a public stand on the event.

Three years down the road, the aspirations of the organizers of the March for Science in 2017 to activate a longer-lasting movement have not been fulfilled, if the declining number of participants at the Marches for Science in 2018 and 2019 is any measure to go by. The public debate on issues that drove demonstrators to the streets at the March for Science in 2017, however, has intensified and is increasingly associated with political developments in Western Europe and North America. The decline of public trust in science, scientific authority and expertise are regarded as enabling features of the rise in conservative populism, white supremacy and attacks on international geopolitical institutions and alliances such as the United Nations and the European Union.

The March for Science has not received attention in STS, but STS scholars have debated the broader issues of the decline of trust in science and the rise of 'post-truth' politics. In 2017, the distinguished STS journal *Social Studies of Science* published a series of conversations on these issues. The conversations were prompted by questions about STS's own implication in these issues. STS was essentially 'blamed for contributing to the decline in trust, by painting a picture in which all facts become claims and all claims are seen as merely political' (Jasanoff & Simmet, 2017: 752). The debate in STS, accordingly, centred around the question of STS's own complicity in disenchanting science and enabling 'post-truth' politics.

One of the articles in this conversation refers to the slogans used at the March for Science in 2017 – 'science is real', 'reality is not up for debate' – to take a stand on this question. The authors consider the simple framing of 'post-truth' politics to be flawed and favour a more differentiated picture in which knowledge and norms are co-produced in political contexts. Accordingly, they argue that scientists need to actively engage their scientific views with competing political positions (Jasanoff & Simmet, 2017: 763).

> To say that facts speak for themselves is to live in a 'post-value' world that ignores contention and questioning as the very stuff of a democracy that has always connected public facts with public values. Reality, indeed, should be up for debate, if that debate is about whose reality counts and by what measures. Avoiding negotiation between facts and values will only result in the blind subjugation of some values over others, with those whose values are left out rejecting the other side's 'truth' as merely politics by another name (Jasanoff & Simmet, 2017: 763).

The example shows that the debate in STS on the contemporary decline of public trust in science and 'post-truth' politics pertains directly to the March for Science. The general terms of the debate, therefore, are of interest to the questions raised about the March for Science at the beginning of this book: What was this demonstration about? Why did these scientists take to the streets rather than use conventional channels of science advocacy? And why did the event take place in 2017? What lies behind this apparent loss of confidence? Why is scientific knowledge seen to be losing authority?

The recent discussion in *Social Studies of Science* on STS's own liability in the decline of public trust and 'post-truth' politics essentially focused on the symmetry principle and the Strong Programme in SSK. The 'logic of symmetry, and the democratizing of science it spawned' are held responsible for inviting 'the scepticism about experts and other elites that now dominates political debate in the US and elsewhere' (Collins et al., 2017: 580). Although symmetry is treated as a contested concept in the debate, in general terms it is employed as proxy for the co-construction of knowledge and social order and as marker for the social and political impact on science and expertise. Bloor's symmetry principle and the Strong Programme in SSK are attributed a central role in the history of STS because they are considered as part of the revolution that 'cracked the pure crystal of science' (Collins et al., 2017: 581):

> Before SSK it was always and only scientific work that was needed to make scientific truth; after SSK what was once seen as the socially sterilised work of experiment and observation became hard to distinguish from political work. By revealing the continuities between science and politics, science studies opened up the cognitive terrain to those concerned to enhance the impact of democratic politics on science but, in so doing, it opened that terrain for all forms of politics, including populism and that of the radical right wing (Collins et al., 2017: 581).

Some authors affirm STS's position as agent engaged in the process of co-construction of science and politics, and indeed consider this form of engagement a requirement for democratic life (Jasanoff & Simmet, 2017). Another position is to refute any inherent political motive of STS, and assert that STS is 'an academic/scientific discipline aimed at understanding the nature of knowledge' rather than a 'political movement for promoting democracy' (Collins et al., 2017: 584). Others ward off difficulties for STS by referencing semantic inaccuracies in the historical use of the term symmetry (Lynch, 2017) or insist on the distinction between scientific and non-scientific knowledge for the body of work of STS (Sismondo, 2017: 589). The problem at hand, however, the decline in trust in science and rise of post-truth politics, is not debated per se. Instead, STS stops short at discussing its own role in these trends.

The differing positions in the debate, however, agree on the high stakes involved and the gravity of the situation. Fuller reminds the STS community that 'symmetry [applies] not only to the range of objects studied but also the range of agents studying them' (Fuller, 2016), implying that STS's own expertise is on trial in the crisis of science and post-truth politics. Collins et al., too, contend that if 'STS cannot find a better way to say why science matters, then STS will be intellectually bankrupt' (Collins et al., 2017: 584). To address this predicament of STS, Collins et al., for example, propose that STS 'find a way to justify expertise in general and scientific expertise in particular' and recommend 'the Studies of Expertise and Experience (SSE), the third wave and elective modernism as programmes to address this dilemma in STS' (Collins et al., 2017: 584). Sismondo advises to build on the achievements of STS and to continue to show how epistemic authority is established and employed (Sismondo, 2017: 589).

The debate clarifies positions but offers no answers to the questions raised about the March for Science and to the resolution of STS's dilemma. In the halls of academia, STS continues to claim a position of unique and objective expertise with regard to matters concerning science, technology and society. At the same time, its own intellectual programme rejects the ideal dissociation from political influences for other areas of science. The symmetry principle and the Strong Programme in SSK, which in many ways paved the way to the establishment of STS as an interdisciplinary academic field in its own right in the 1980s, now threatens to undermine this field's very legitimacy in the 'post-truth' era. Of interest here is the resurgence of the symmetry principle and the Strong Programme in SSK in connection with the debate on STS's intellectual and analytical exclusivity.

The recent debate in STS on questions about the connection between 'symmetry' in Science and Technology Studies (STS) and 'post-truth' politics is relevant to the concerns that drove scientists to participate in the March for Science. The debate links the interpretation of the March for Science to the symmetry principle in STS and to the Strong Programme in the SSK. In a sense, the scientists' slogans may be read as an attempt to refute Bloor's contentions: they rebut the *causal* effect of external influences on science; they reject an *impartial* perspective on truth and falsity, and on rationality or irrationality; and they defend the need for *asymmetrical* explanations for true and false beliefs.

The debate in STS is also relevant to the March for Science when considering the question of why this academic community did not seize upon the historical event to offer its expertise. The silence could be explained by STS's predicament: on the one hand, the STS scholar may be tempted to side with the scientists and experts of the March for Science to insist on standards that separate his or her own scientific work from non-scientific opinions. On the other hand, the STS researcher is unable to relinquish the symmetry principle and the Strong Programme in SSK, because they are hallmarks of STS's intellectual history and confer academic authority and identity on its community.

STS's conversation on the decline of public trust in science and 'post-truth' politics suggests that the terms of this debate remain confined to the conceptual boundaries of the Strong Programme in SSK. This book argues that postcolonial perspectives offer a path out of this stalemate situation and can shift the terms of the debate on the crisis of science and society beyond these boundaries. A brief postcolonial perspective on the March for Science is offered below to showcase this move.

7.2 A postcolonial perspective on the March for Science

Perhaps the most inspiring aspect of the March for Science, and what may prove to be its most enduring legacy, is its truly global nature. Science is not western; it is everywhere and for everyone.
Mark Lynas, *The Guardian*, 18 April 2017.

The March for Science originated in communications on Reddit, a social media site self-described as the 'front page of the internet', according to its organizers. The early idea was for scientists to demonstrate in Washington D.C., the centre of political power in the US. The proposal quickly spread through social media platforms such as Facebook and Twitter, prompting others to organize local marches. An internet website was created to link local marches within and across countries to mobilize for the event. In this narrative, the March for Science simply spread randomly and democratically by way of the internet and quickly achieved global status by virtue of its geographic distribution in more than five hundred cities.

The postcolonial perspective warrants closer examination of the global claims in the narrative of the March for Science. It calls for a more detailed appraisal of the geographical distribution of the Marches for Science, of the affiliations of their organizers, of the procedures involved in disseminating the idea, and of the particular images of science that were endorsed by the event. Even a quick look at these issues is sufficient to demonstrate the value of the postcolonial perspective.

The geographical distribution of the Marches for Science in over five hundred cities was highly concentrated in Western Europe and North America (see Figure 1). For example, a total of nineteen marches were organized in Germany, but only one single march each took place in all of Russia, India and China. No marches occurred in Northern African countries and the

Middle East. On the other hand, marches were carried out in Uganda, Nigeria and Ghana. How can this skewed distribution be explained? And in what sense does it support the claim for a global scope of the March for Science?

At first glance, the researcher might be tempted to suggest that the pattern of distribution of the Marches for Science follows the global distribution of scientific proficiency and expertise more generally. The clustering of marches in North America and Western Europe, in this view, might be seen to express a broader global knowledge divide between developed countries on the one side, and developing and emerging countries on the other side. In this picture, the global distribution of marches could be explained in the terms of Bloor's tenets, which would attribute different beliefs or states of knowledge to the geographic locations on both sides of this divide. To investigate the divide, the researcher could study its formative causal conditions using the same types of cause to explain true and false beliefs, science and non-science, knowledge and non-knowledge. In the end, an impartial and symmetrical style of explanation could be given in the different claims for knowledge on either side of the divide (Bloor, 1991 [1976]).

At second glance, however, the pattern of distribution of the Marches for Science in 2017 turns out to be more complex, and escapes the explanatory power of Bloor's tenets. It does not accord with the patterns of global scientific proficiency and expertise of the time. The global political economy of science in the twenty-first century has undergone profound changes. The scientific stronghold of Western Europe, North America, and Japan of the last century has given way to an increasingly 'multi-polar' constellation in which emerging and developing economies assume new roles. The archetypal image of a simple knowledge divide is no longer defensible and the global political economy of knowledge has become more intricate. As an example, even though only one March for Science was organized in China in 2017, this country accounted for almost twenty percent of global research expenditure, reaching parity with Europe in that year.

Furthermore, the analytical gaze of Bloor's tenets also would fail to explain why marches were carried out in certain African countries, such as Uganda, Nigeria and Ghana, and not in others. How did the organizers of the Marches in Uganda, Nigeria and Ghana receive notice of the event? Why did other African countries, such as Egypt, Ethiopia, and Morocco, not participate in the March for Science?

A quick look at the websites on the Marches for Science in Uganda, Nigeria and Ghana offers indications on the reasons why marches were organized in these countries. Their websites all link back to a non-profit institution, the *Cornell Alliance for Science*, associated to Cornell University, a North American Ivy League University. An official partner to the March for Science, the Cornell Alliance for Science reported in 2017 that it had mobilized its global network to assist in transforming the historical action of the March for Science into an international affair. Explicit mention is made of the countries Kenya, Nigeria, Ghana, Uganda, Mexico, Hawaii, the Philippines and Bangladesh. The organization actively mobilized for the March for Science in African countries, and internationally. Its mission, purpose, activities and funding sources might shed some light on the motives for taking this initiative.

The mission of the Cornell Alliance for Science is to build 'a global network of science allies' by offering training courses to 'equip and empower emerging international leaders who are committed to advocating for science-based communications and access to scientific innovation in their home countries'. Although its name implies a broader purpose, the organization was originally launched in 2014 as an effort to 'depolarize the charged debate' around genetically modified foods (GMOs) (Malcan, 2019). The sustained focus of the alliance on GMOs is evident from in its training courses which include strategic planning, grassroots organizing and communications around agricultural biotechnology. The particular focus on Africa transpires from resources that are offered on its websites which aim to 'debunk' the 'most common myths' on GMOs, which are attributed to 'a massive disinformation campaign' on agricultural biotechnology. The alliance maintains close ties to agricultural and chemical industries that advocate access to genetically engineered crops. The primary source of funding for the Cornell Alliance for Science stems from the Bill & Melinda Gates Foundation, with further grants and donations from other institutions and individuals. Incidentally, the author of the lead article on the March for Science in the British newspaper *The Guardian* on 17 April 2017, who was quoted at the beginning of this section and in the first paragraph of this book ('Scientists to take to the streets in global march for truth') has been sponsored as visiting fellow by the Alliance.

The purpose of this brief profile is not to call into question the motives of the Cornell Alliance for Science in advocating for the March for Science. The example is of interest because the mere existence of the connections between a non-profit organization associated with a North American university, the Marches for Science in Kenya, Nigeria, Ghana, Uganda, Mexico, Hawaii, the Philippines and Bangladesh, and an article on the 'Global March for Truth' in a widely-read British newspaper, disproves the narrative of randomness that the organizers claim for the spread of the March for Science across the world. It also raises questions as to the objectives and purposes of these Marches for Science as proclaimed by their organizers. Instead, the connections in this example suggest complex layers of purposes and dependences that are eclipsed in the mainstream storyline of the March for Science.

The example shows that the spread of the idea of the March for Science was not arbitrary and unplanned and did not simply proliferate across the globe on the basis of the merit of the idea and its global relevance. The March for Science did not qualify as a global occasion because its concerns were inherently global in nature but because those who stood to judge on the scope of the March considered their concerns and claims to be universal. The pattern of distribution of Marches by no account stands the test of being global; by far the greater number of Marches took place in Western Europe and North America, the location of the renowned institutions of science who endorsed and issued public statements on the march, the leaders and employees who gave interviews and wrote opinion pieces, and the media outlets that reported on the event. Nevertheless, their endorsement was described as a random, undirected, global process that occurred without any advocacy activities of interest groups.

The process and implicit criteria by which the March for Science was staged as a global occasion, warrant closer investigation because they offer indications on the particular image of science that inspired the event. This image is important because it informs the diagnosis of a crisis between science and society as a decline of public trust in science and the rise of 'post-truth' politics. The postcolonial perspective draws attention to new questions about the March for Science that challenge the alleged arbitrariness of its global spread and call for a closer look at the storyline. How did the idea of a March for Science come to be staged as a global concern? What particular images of science and knowledge were promoted by the event? To what extent can the current crisis of science and society, as identified to be one of declining public trust in science and 'post-truth' politics, be considered a global phenomenon?

With these questions, the postcolonial perspective adds a new interpretive layer to the cause of the demonstrating scientists at the March for Science. Rather than simply protesting to restore a separation of truth and power, of scientific expertise and 'post-truth' politics, they also took to the streets to defend a particular image of science, its institutions and procedures of evaluation, its traditions to determine criteria for expertise and standards. The postcolonial perspective also sheds new light on the scope of the 'historically unprecedented' political movement of scientists in the March for Science. The skewed global participation in the March for Science suggests that the concerns of the protesting scientists were, in the first place, directed at audiences in North America and Europe. The alleged global scope of the event furnished the scientists with a universal cause and professional vocation in local settings. By implication, the crisis in the relationship between science and society that motivated the March for Science, perceived as a decline of public trust in science and 'post-truth' politics, loses its validity as a global diagnosis. The postcolonial perspective shifts the terms of the debate on the crisis between science and society. The terms by which STS has been debating this crisis may not be suited to the new scope of questions.

7.3 Conclusions

Certainly STS has work to do to explain why the Enlightenment project has taken a hit in recent years; and like any social science discipline with a stake in progress, STS should consider how its perspectives can lead us forward from this moment of anxiety and popular disenchantment, not least by helping to better diagnose our present predicament.
Jasanoff and Simmet, 2017.

The March for Science, held in over five hundred cities across the world, was described as a global manifestation of the public decline of trust in science and the rise of 'post-truth' politics. These matters have been debated in STS as a crisis in the relationship between science and society. The debates in STS have focused on this field's own role in the crisis and have revolved around the symmetry principle and the Strong Programme in SSK. The stakes are high and involve nothing less than STS's own legitimacy as an interdisciplinary academic field of knowledge in its own right. However, the debate about the implications of its own analytical principles on STS appears to have reached gridlock. STS's silence on an unprecedented event in the history of science, where hundreds of thousands of scientists took to the streets to demonstrate in a March for Science, supports this presumption. This performance raises serious questions with regard to the conceptual equipment of STS to read, analyse and interpret the dynamics of the contemporary relationship between science and society.

The postcolonial perspective on the March for Science has shown that the terms of the debate in STS on the public decline of trust in science and 'post-truth' politics have limited analytical scope to capture some of the key characteristics of this historically unique event. It dismantles the claim that the March for Science simply spread across the globe randomly and by virtue of the universal relevance of its cause. Instead, postcolonial considerations reveal that active forces were at work to spread the idea of the March for Science. These forces advocated particular images and standards of science. In this way, the March for Science stages science as a *North Atlantic universal* and enacts the transition narrative by advocating disembodied claims for knowledge with particular standards and characteristics. The four tenets of the Programme in Science Studies Elsewhere uncover the *entangled, relational, uneven* and *contestable* disposition of the March of Science.

The postcolonial perspective is not a newcomer to STS but for many years has formed part of the heterogeneous set of approaches to the study of science, technology and society that are the hallmark of this interdisciplinary field. Why, then, not simply undertake a thorough empirical analysis of the March for Science from a postcolonial perspective to contribute to the debate in STS? In any case, would not the natural expansion of STS communities to postcolonial contexts in one way or another eventually lead to pluralized perspectives and newly situated knowledge? Why embark on a new Programme in Science Studies Elsewhere? The record suggests that postcolonial empirical research has not succeeded in suspending the Collective Blind Spot of STS, which prevents the debate from moving forward. Mainstream STS has warded off the implications of postcolonial thought for its 'epistemic land-scape' (Felt et al., 2017) by replicating the transition narrative of science and technology through its analytical tools and concepts.

Thomas P. Hughes replicates the transition narrative of technology when he develops a theory of technological change based on three cities in 'Western society', Chicago, Berlin and London. The transition narrative becomes effective when his historiography weaves a seamless web of technology and society that passes through the historical phases of invention and development, to technology transfer, to system growth and maturity. This historiography enacts the transition narrative by affording technological primacy to the idea of Western society. To be sure, Hughes' sequence of events is easily disproved by the historical record. The idea of regional electric power systems was not put to test in Berlin and London and transported from there to the rest of the world. Instead, Johannesburg was used as an experimental foil to test the idea of large electric power systems before the regional systems of Berlin and London were constructed.

Notably, the circumstances for constructing a regional electric power system in Johannesburg differed considerably from those in Berlin and London. The gigantic power system of Johannesburg was built before the First World War to provide electric power to the gold mining industry – not to the young city of Johannesburg. It was financed, built and owned by the imperial powers of Germany and Great Britain and used as a textbook prototype to advocate for large regional electric power systems in European cities. When Hughes' model of technological change disregards the influence of settings outside of the geographical boundaries of 'Western society' for the history of electrification in Berlin and London, it enacts the transition narrative. As soon as STS scholars employ Hughes' framework and concepts to study technological change, this mechanism sets in and is silently replicated.

The enactment and replication of the transition narrative of science and technology in STS's analytical concepts and approaches are not simply a matter of intellectual concern. Disregard of the prototype role of the large electric power system in Johannesburg for the historiography of electrification leads to the omission of key historical drivers and agents, and falsifies the role of others. For example, the concept of *society* in Hughes' seamless web makes no sense as an analytical category in the context in which the gigantic power system of the Witwatersrand was built. The population of Johannesburg had grown in tandem with the mushrooming gold mining industry. The historical record presents a complex assemblage of inhabitants who escaped univocal categorization. Various indigenous peoples from Southern Africa and adjacent colonies (e.g. Portuguese), Asian migrants, European and American immigrants, Boer and British settlers, and others, populated Johannesburg. Rather than forming a society in the sense of Hughes' concept, these populations were organized into a hierarchical labour administration that was based on the idea of racial supremacy and used categories such as 'kafirs', 'natives', 'black labourers', 'coloured labourers', or 'white labourers'. Hughes' idea of a seamless web of technology and society simply does not apply to these circumstances. Instead, the transition narrative of his history of technology settles as a Blind Spot on his theory and concepts on technological change and is silently replicated in STS.

This book proposes a postcolonial perspective on science and technology as a means to deal with the dilemma of STS's Collective Blind Spot. The spot has not budged in more than thirty years of empirical postcolonial STS studies. Empirical postcolonial analysis alone is unlikely to succeed in contributing to the critical questions STS is currently grappling with. Neither is the Collective Blind Spot likely to simply disappear with an expansion of the STS community to new geographical sites. The postcolonial analytical approach is articulated in the Programme in Science Studies Elsewhere to shift the terms of the debate in which STS currently appears to be trapped.

The epistemic landscape of the latest *Handbook of Science and Technology Studies* (2017) confirms the need for a programmatic approach to escape the dilemma of the Collective Blind Spot in STS. The *Handbook* emphasizes the intellectual commitment of STS to the tradition of studying situated knowledge claims (Haraway, 1988) and the political geography in 'the making of knowledge' (Felt et al., 2017: 13, 14). But it does not translate the commitment into the epistemic landscape of the *Handbook* and as a result only pays lip service to the problem. The *Handbook* acknowledges in general terms the influence of locations, experiences and traditions on the valuation regimes and political economy of STS, and recognizes concomitant obligations and responsibilities (Felt et al., 2017). But it does not shoulder this responsibility when it selects topics and authors for the chapters to set standards and frame problems for the field of STS. The *Handbook* rejects the assumption of an 'uneven distribution of who gets voice' in the *Handbook* and accentuates the field's expansion on the international level, to new associations in East Asia, Japan, and Latin America or to new locations for STS conferences in Tokyo (2010) and Buenos Aires (2014) (Felt et al., 2017: 16). But it substantiates this assumption when the voices and intellectual work of this international community are not represented in its content, and roughly 90 % of chapters are written by authors affiliated with institutions in North America and Europe.

The *Handbook* does not translate its reasoning into action. It sidesteps the consequences of STS's own intellectual analytical tools on its epistemic landscape and perpetuates the Collective Blind Spot of STS. This status-quo demonstrates that STS's strategy of awaiting a steady accumulation of diasporic, situated knowledges into the intellectual fabric has not been successful.

This book proposes the Programme in Science Studies Elsewhere to shift the terms of the debate in STS. This shift requires attention to be given to the geographies of management and of imagination that underlie science and technology as *North Atlantic universals.* These analytical lenses are easily looked down upon because they are ambiguous and difficult to grasp – yet their importance in shaping science and technology is evident. Trouillot's notion of *Elsewhere* reveals mechanisms in the political economy of knowledge and science, but this analysis does not run along the lines of geopolitical or national boundaries. It follows the contours of the geography of management and the geography of imagination. This is a complicating, abstract step that distinguishes Trouillot's *Elsewhere* from other notions of the Other. In Trouillot's geographies, geopolitical boundaries play a secondary role. In other words, his cartography of *Elsewhere* foreshadows the geopolitical analytical lens: Had Shapin and Schaffer been Ghanaian and Chinese rather than American and British STS scholars writing the book at a university in Buenos Aires, these circumstances would not have changed the book's reproduction of the transition narrative of science. Put differently, the strategies and mechanisms of replication in STS do not fall under any exclusive proprietary rights within the geopolitical boundaries of Euro-America, but abide by the maps of the geographies of management and imagination (Hofmänner, 2016).

Thomas A. Edison *imagined* a global empire of rights and privileges for his electric light and power station project before he developed the technology. He secured these rights and privileges in the form of patents and companies across the globe before the technical and economic feasibility of the technology had been ascertained. His Edison Indian and Colonial Electric Company Ltd. formed part of the imperial wrangle over territorial rights and privileges in colonial settings. Edison's imaginary geography of far-away places had a bearing on the design and the success of his electric light and electric power station technology.

The German engineer Georg Klingenberg *imagined* a system of large power stations to provide electricity to the gold mining companies of Johannesburg at the beginning of the twentieth century, barely two decades after gold had been discovered in the area. This grand system was to be financed and owned by the imperial powers Germany and Great Britain, who incidentally were also heavily invested in the gold mining business in Johannesburg. Once constructed, this prototype informed Klingenberg's educational book on '*Large Electric Power Stations. Their Design and Construction*' (1916) which influenced the imagination of an entire generation of power station engineers. Thomas Hughes referenced Klingenberg in *Networks of Power* as 'an engineer of international reputation and authority on power plant design and operation' (Hughes, 1983: 197). But the seamless web of technology and society *imagined* by Hughes prevented him from discerning the international entanglements of Klingenberg's activities because his notion of 'Western society' was contrasted against its absence, *Elsewhere*.

The experimental philosophers Robert Boyle and Thomas Hobbes *imagined* 'inferior creatures' and 'savages' respectively, as oppositional referents to their philosophical programmes. These referents played a role in shaping the model profile of the experimental natural philosopher and his role in the beginnings of modern science, an historical episode widely referred to as the origins of modernity. In this way, the cultural heritage to experimental practice becomes a standard for science that is contrasted to its absence, *Elsewhere*.

As pioneers in the field of STS, the scholars Steven Shapin and Simon Schaffer *imagined* 'the ignorant stranger' to experimental life as a method for undertaking a symmetrical study of accepted and rejected knowledge in 17th century Restoration England. Their study concluded that experimental practice delivered a solution to the problems of knowledge and social order. This conclusion has influenced the course of STS and continues to be cited frequently by its scholarly community. However, it overlooked that experimental practice as a solution to the problems of knowledge and social order at the same time created a solution to the problems of non-knowledge and social disorderliness, *Elsewhere*.

STS scholars have *imagined* the field of STS as an increasingly international and diverse field of study. Yet, the record shows that the intellectual landscape and analytical scope of this field continues to be defined by a handful of countries of the North Atlantic. This contradiction is sustained by STS's Collective Blind Spot, and its replication of *Elsewhere* in the study of modern science and technology.

The protesters at the March for Science *imagined* a 'global' event, despite the evident clustering of marches in Western Europe and North America, and notwithstanding the low uptake of the march outside these regions. They *imagined* an international crisis in the relationship between science and society that is characterized by the decline of public trust in science and the rise of 'post-truth' politics. The postcolonial perspective on the March for Science suggests that this characterization may not adequately capture the range of forces at play in shaping this relationship across the globe. The postcolonial history of science and technology bears testimony to the power of imagination in shaping the global political economy of knowledge. The Programme in Science Studies Elsewhere directs STS's attention to these forces that transform knowledge, power and truth across the globe.

This book ends by returning to the scenery visible from the viewing deck on the fiftieth floor of the Carlton Centre in Johannesburg on 22 April 2017. Unlike sightseers in Chicago, Berlin and London, the visitor to the 'Top of Africa' skyscraper would not have seen any protesters marching for science in the streets of the City of Gold because no March for Science took place in Johannesburg on this day.

Abbreviations

A.E.G.   Allgemeine Elektrizitätsgesellschaft, Berlin
ANT   Actor Network Theory
BSAC   British South Africa Company
CSSR   Committee on Science and its Social Relations
EASST   European Association for the Study of Science and Technology
EASTS   East Asian Science, Technology and Society: An International Journal
EC   European Commission
ECB   Electricity Control Board
EPOR   Empirical Programme of Relativism
Escom   Electricity Supply Commission
GEPC   General Electric Power Company Ltd.
ICSU   International Council of Scientific Unions
ISCOR   South African Iron and Steel Corporation
LTS   Large Technological Systems
RCEW   Rand Central Electric Works Ltd.
SCOT   Social Construction of Technology
STS   Science and Technology Studies
VFPC   Victoria Falls Power Company
VFTPC   Victoria Falls and Transvaal Power Company
4S   Society for Social Studies of Science

Tables

Table 1 Key questions for the empirical study of the North Atlantic universals with the lenses of the geographies of management and imagination, (based on Trouillot, 1991, 2002).

Table 2 The five phases in the development of electrical power systems in Hughes' *Networks of Power: Electrification in Western Society, 1880–1930.*

Table 3 Specifications on the Rand Central Electric Works Ltd. (RCEW) and the General Electric Power Company Ltd. (GEPC).

Table 4 Total capacity of power installed and in progress in the four power stations of Johannesburg in 1913 (Hadley, 1913).

Table 5 Particulars of electricity consumers (mines, industries, municipalities) supplied by the VFTPC in Johannesburg in 1915 (Price 1915: vi).

Table 6 The total electricity generating capacity in South Africa in 1915 (supplied by the VFTPC) and in 1930 (supplied by the VFTPC and Escom) (Troost & Norman, 1969).

Table 7 The Programme in Science Studies Elsewhere in reference to the four tenets of Bloor's Strong Programme in the Sociology of Scientific Knowledge.

Figures

References

Adams, Vincanne. 2002. "Randomized Controlled Crime: Postcolonial Sciences in Alternative Medicine Research". *Social Studies of Science*, Vol. 32, No. 5/6, pp. 659–690.

Albrecht, Henning. 2011. *Alfred Beit. Hamburger und Diamantenkönig.* Hamburg: Hamburgische Wissenschaftliche Stiftung.

Anderson, Warwick. 2002. "Introduction: Postcolonial Technoscience". *Social Studies of Science*, Vol. 32, No. 5/6, pp. 643–658.

Anderson, Warwick. 2007. "How Far Can East Asian STS Go? A Commentary". *East Asian Science, Technology and Society: An International Journal*, Vol. 1, No. 2, pp. 249–250.

Anderson, Warwick and Vincanne Adams. 2008. "Pramoedya's Chickens: Postcolonial Studies of Technoscience". In Edward J. Hackett, Olga Amsterdamska, Michael Lynch, and Judy Wajcman (eds). 2008. *Handbook of Science and Technology Studies*, Third Edition. Cambridge: MIT Press, pp. 181–204.

Appadurai, Arjun. 1996. *Modernity at Large: Cultural Dimensions of Globalization.* Minneapolis: University of Minnesota Press.

Appadurai, Arjun. 1999 [1993]. "Disjuncture and Difference in the Global Cultural Economy". In During, Simon (ed). 1999 [1993]. *Cultural Studies Reader*, Second edition. New York: Routledge, pp. 220–232.

Barnes, Barry. 1974. *Scientific Knowledge and Sociological Theory.* Chicago: University of Chicago Press.

Barnes, Barry. 1977. *Interests and the Growth of Knowledge.* London: Routledge.

Bauchspies, Wenda K., Croissant, Jennifer and Sal Resivo. 2006. *Science, Technology, and Society: A Sociological Approach.* London: Wiley-Blackwell.

Beauchamp, Ken G. 1997. *Exhibiting Electricity.* London: The Institution of Electrical Engineers.

Benton, Lauren. 2010. *Search for Sovereignty. Law and Geography in European Empires, 1400–1900.* Cambridge, New York: Cambridge University Press.

Bhabha, Homi K. 1994. *The Location of Culture.* London UK/New York: Routledge.

Bhambra, Gurminder K. 2007. *Rethinking Modernity: Postcolonialism and the Sociological Imagination.* New York: Palgrave Macmillan.

Bhambra, Gurminder K. 2011. "Historical Sociology, Modernity, and Postcolonial Critique". *American Historical Review*, Vol. 116, No. 3, pp. 653–662.

Biagioli, Mario (ed). 1999. *The Science Studies Reader*. New York: Routledge.

Biagioli, Mario. 1999. "Introduction: Science Studies and its Disciplinary Predicament". In Biagioli, Mario (ed). 1999. *The Science Studies Reader*. New York: Routledge, pp. xi–xviii.

Bijker, Wiebe E. 2010. "How is technology made?—That is the question!" *Cambridge Journal of Economics*, Vol. 34, pp. 63–76.

Bijker, Wiebe E. and Trevor J. Pinch. 2012. "Preface to the Anniversary Edition". In Bijker, W. E., Hughes, T. P. and T. J. Pinch (eds) 2012 [1987]. *The Social Construction of Technological Systems: New Directions in the Sociology and History of Technology*. Cambridge, Mass.: MIT Press. Anniversary Edition, pp. xi – xxxiv.

Bijker, Wiebe E., Hughes, Thomas P. and Trevor J. Pinch (eds). 2012 [1987]. *The Social Construction of Technological Systems: New Directions in the Sociology and History of Technology*. Cambridge, Mass.: MIT Press. Anniversary Edition.

Bloor, David. 1991 [1976]. *Knowledge and Social Imagery*. Second edition. Chicago: University of Chicago Press.

Bloor, David. 1999. "Anti Latour". *Studies in the History and Philosophy of Science*, Vol. 30, No. 1, pp. 81–112.

Bloor, David. 1999. "Reponse to Latour". *Studies in the History and Philosophy of Science*, Vol. 30, No.1, pp. 131–136.

Boatcă, Manuela, Costa, Sérgio, and Encarnación G. Rodríguez. 2010. *Decolonizing European Sociology: Interdisciplinary Approaches*. Farnham: Ashgate.

Boatcă, Manuela, Costa, Sérgio, and Encarnación G. Rodríguez. 2010. "Introduction: Decolonizing European Sociology. Different Paths towards a Pending Project". In Boatcă, M., Costa, S., and Rodríguez, E. G. (eds). 2010. *Decolonizing European Sociology. Interdisciplinary Approaches*. Farnham: Ashgate, pp. 1–12.

Böhm, Ekkehard. 1973. Hamburger Grosskaufleute in Südafrika zu Ende des 19. Jahrhunderts. *Zeitschrift des Vereins für Hamburgische Geschichte*, Band 59. Hamburg, Christians.

Bright, Edward Brailsford and Charles Bright. 2012 [1898]. *The Life Story of the late Sir Charles Tilston Bright, Civil Engineer. With Which is Incorporated the Story of the Atlantic Cable, and the First Telegraph to India and the Colonies*. Volume 2. New York, Cambridge University Press.

Bright, Charles and D. E. Hughes. 1898. Report Upon the International Exhibition of Electricity in Paris, 1881. In Bright, Edward Brailsford and Charles Bright. 2012 [1898]. *The Life Story of the late Sir Charles Tilston Bright, Civil Engineer. With Which is Incorporated the Story of the Atlantic Cable, and the First Telegraph to India and the Colonies*. Volume 2. New York, Cambridge University Press, pp. 590–601.

Bruun, Henrik and Janne Hukkinen. 2003. "Crossing Boundaries. An Integrative Framework for Studying Technological Change". *Social Studies of Science*, Vol. 33, No. 1, pp. 95–116.

Bucchi, Massimiano. 2004. *Science in Society: An Introduction to Social Studies of Science*. London: Routledge.

Busch, Lawrence. 1987. Review: Leviathan and the Air-Pump: Hobbes, Boyle, and the Experimental Life by Steven Shapin, Simon Schaffer. *Science & Technology Studies*, Vol. 5, No. 1 (Spring, 1987), pp. 39–40.

Cackowski, Zdzisław. 1982. "Ludwik Fleck's Epistemology". *Dialectics and Humanism*, No. 3, pp. 11–23.

Campion, Patricia and Wesley Shrum. 2004. "Gender and Science in Developing Areas". *Science, Technology, and Human Values*, Vol. 29, No. 4, pp. 459–485.

Carpenter, Frank G. 1908. "Harnessing the Victoria Falls. Industrial Revolution Likely to Follow the Utilisation of the Tremendous Force". *Chicago Sunday Tribune*, 28 November 1908.

Casson, Mark. 1994. "Institutional Diversity in Overseas Enterprise: Explaining the Free-Standing Company". *Business History*, Vol. 36, No. 4, pp. 95–108.

Chakrabarty Dipesh. 1992. "Postcoloniality and the Artifice of History: Who Speaks for 'Indian' Pasts?". *Representations*, Special Issue: Imperial Fantasies and Postcolonial Histories (Winter, 1992), No. 37, pp. 1–26.

Chakrabarty, Dipesh. 2002a. "Europa provinzialisieren. Postkolonialität und die Kritik der Geschichte". In Conrad, S. and Randeria, S. (eds). 2002. *Jenseits des Eurozentrismus: Postkoloniale Perspektiven in den Geschichts- und Kulturwissenschaften*. Frankfurt a. M.: Campus Verlag, pp. 283–312.

Chakrabarty, Dipesh. 2002b. *Habitations of Modernity*. Chicago: University of Chicago Press.

Chakrabarty Dipesh. 2008 [2000]. *Provincializing Europe: Postcolonial Thought and Historical Difference*. Princeton: Princeton University Press.

Chambers, David Wade and Richard Gillespie. 2000. "Locality in the History of Science: Colonial Science, Technoscience, and Indigenous Knowledge". *Osiris*, Vol. 15, pp. 221–240.

Chikowero, Moses. 2007. "Subalternating Currents. Electrification and Power Politics in Bulawayo, Colonial Zimbabwe, 1894–1939". *Journal of Southern African Studies*, Vol. 33, No. 2, pp. 287–306.

Christie, Renfrew. 1984. *Electricity, Industry, and Class in South Africa*. Albany (NY): State University of New York Press.

Christopher, A. J. 1988. "Roots of urban segregation: South Africa at Union 1910". *Journal of Historical Geography*, Vol. 14, pp. 151–169.

Cohen, Bernard I. 1987. Review: Leviathan and the Air-Pump: Hobbes, Boyle, and the Experimental Life by Steven Shapin, Simon Schaffer. *The American Historical Review*, Vol. 92, No. 3 (Jun., 1987), pp. 658–659.

Cohen, Robert S. and Thomas Schnelle (eds). 1986. *Cognition and Fact. Materials on Ludwik Fleck.* Boston Studies in the Philosophy of Science, Volume 87. Dordrecht: D. Reidel.

Collins, Harry M. and Robert Evans. 2002. "The Third Wave of Science Studies: Studies of Expertise and Experience". *Social Studies of Science*, Vol. 32, No. 2, pp. 235–296.

Collins, Harry M., Evans, Robert and Martin Weinel. 2017. "STS as science or politics?" *Social Studies of Science*, Vol. 47, No. 4, pp. 580–586.

Conrad, Sebastian, Randeria, Shalini and Beate Sutterlüty. 2002. *Jenseits des Eurozentrismus: Postkoloniale Perspektiven in den Geschichts- und Kulturwissenschaften.* Frankfurt a.M.: Campus Verlag.

Coppersmith, Jonathan. 2004. "When Worlds Collide: Government and Electrification, 1892–1939". *Business History Conference*, pp. 1–31.

Corbey, Raymond and Joseph Leerssen (eds). 1991a. *Alterity, Identity, Image. Selves and Others in Society and Scholarship.* Amsterdam: Rodopi.

Corbey, Raymond and Joseph Leerssen. 1991b. "Studying Alterity: Backgrounds and Perspectives". In Corbey, R. and J. Leerssen (eds). 1991a. *Alterity, Identity, Image, Selves and Others in Society and Scholarship.* Amsterdam: Rodopi, pp. vi–xviii.

Costa, Sérgio. 2005. "Postkoloniale Studien und Soziologie: Differenzen und Konvergenzen". *Berliner Journal für Soziologie*, Vol. 15, No. 2, pp. 283–294.

Craig, Gordon, A. 1978. *Germany, 1866–1945.* Oxford (NY): Oxford University Press.

Cutcliffe, Stephen H. 1989. "Science, Technology, and Society Studies as an Interdisciplinary Academic Field". *Technology in Society*, Vol. 11, pp. 419–425.

Cutcliffe, Stephen H. 2000. *Ideas, Machines, and Values: An Introduction to Science, Technology, and Society Studies.* Oxford: Rowman & Littlefield Publishers.

de Laet, Marianne and Annemarie Mol. 2000. "The Zimbabwe Bush Pump: Mechanics of a fluid technology." *Social Studies of Science*, Vol. 30, No. 2, pp. 225–263.

Dernburg, Bernhard. 1907. *Zielpunkte des Deutschen Kolonialwesens.* Zwei Vorträge. Berlin: Mittler.

Draper, Ralph J. 1967. *The Engineer's Contribution. A History of the South African Institute of Mechanical Engineers 1892–1967.* Johannesburg: Kelvin House.

Edge, David. 1995. "Reinventing the Wheel". In Jasanoff, S., Markle, G. E., Peterson, J.C. and Pinch T. (eds). 1995. *Handbook of Science and Technology Studies.* London: Sage, pp. 3–27.

Edison, Thomas Alva. 1883. *The Edison System of Incandescent Electric Lighting, as Applied in Mills, Steamships, Hotels, Theatres, Residences, &c. by the Edison Company of Isolated Lighting.* 65 Fifth Avenue, New York City, 1883. New York: C.O. Burgoyne Printer.

Elkana, Yehuda. 2000: *Rethinking – Not Unthinking – the Enlightenment. Debates on Issues of Our Common Future.* Weilerswist: Velbruck Wissenschaft.

Elkana, Yehuda, Krastev, Ivan, Macamo, Elisio, and Shalini Randeria (eds). 2002. *Unravelling Ties: From Social Cohesion to New Practices of Connectedness.* Frankfurt a.M./ New York: Campus/St. Martin's Press.

Elzinga, Aant. 1997. "Some Notes From the Past". Originally published in the *EASST Review*, Vol. 14, No. 2 (1997). http://easst.net/wp-content/uploads/2012/04/Some-Notes-From-EASSTs-Past.pdf. (accessed 19 November 2012)

Elzinga, Aant. 2004. "The New Production of Reductionism in Models Relating to Research Policy". In Grandin, K., Wormbs, N. and S. Widmalm (eds). 2004. *The Science–Industry Nexus: History, Policy, Implications.* Science History Publications/ USA and the Nobel Foundation. Sagamore Beach MA: Watson Publishing International, pp. 277–304.

Elzinga, Aant. 2012. "Features of the current science policy regime: Viewed in historical perspective". *Science and Public Policy*, Vol. 39, No. 4, pp. 416–428.

Elzinga, Aant and Andrew Jamison. 1995. "Changing Policy Agendas in Science and Technology". In S. Jasanoff, G. Markle, J. Petersen and T. Pinch (eds). 1995. *Handbook of Science and Technology Studies.* Thousand Oaks, CA: Sage, pp. 572–597.

Engels, Anita and Tina Ruschenburg. 2008. "The uneven spread of global science: patterns of international collaboration in global environmental change research". *Science and Public Policy*, Vol. 35, No. 5, pp. 347–360.

Escobar, Arturo. 1994. *Encountering Development. The Making and Unmaking of the Third World.* Princeton: Princeton University Press.

European Commission (EC). 2008. "A Strategic Framework for International Science and Technological Co-operation". COM 2008/588 final.

European Commission (EC). 2012. "Enhancing and focusing EU international co-operation in research and innovation: A strategic approach". Communication from the Commission to the European Parliament, the Council, the European Economic and Social Committee and the Committee of the Regions. Brussels, 14 Sept. 2012.

Fan, Fa-Ti. 2007. "East Asian STS: Fox or Hedgehog?" *East Asian Science, Technology and Society: An International Journal*, Vol. 1, No. 2, pp. 243–247.

Feingold, Mordechai. 1991. Review: Leviathan and the Air–Pump. Hobbes, Boyle, and the Experimental Life by Steven Shapin, Simon Schaffer. *The English Historical Review*, Vol. 106, No. 418 (Jan., 1991), pp. 187–188.

Felt, Ulrike, Nowotny, Helga and Klaus Taschwer. 1995. *Wissenschaftsforschung. Eine Einführung.* Frankfurt am Main: Campus Verlag.

Felt, Ulrike, Fouché, Rayvon, Miller, Clark A. and Laurel Smith-Doerr. 2017. *The Handbook of Science and Technology Studies.* Fourth Edition. Cambridge, Mass: MIT Press.

Fischer, Peter. 1985. Concessions. In *Encyclopedia of Public International Law*. Amsterdam: Elsevier Science Publishers, pp. 100–106.

Fleck, Ludwik. 1980 [1935]. *Entstehung und Entwicklung einer wissenschaftlichen Tatsache*. Suhrkamp: Frankfurt.

Forbes, B.C. 1917. *Men Who Are Making America*. New York, B.C. Forbes Publishing Co.

Forbes, George. 1898. "Long Distance Transmission of Electric Power". *Journal of the Society of Arts*, Vol. XLVII, No. 2401 (November 1898), pp. 25–40.

Fraser, Maryna. 1975. The Barlow Rand Archives, Johannesburg. *South African Archives*.

Fuller, Steve and James H. Collier. 2004 [1993]. *Philosophy, Rhetoric, and the End of Knowledge. A New Beginning for Science and Technology Studies* (2nd ed.). Mahwah, New Jersey: Lawrence Erlbaum Associates.

Fuller, Steve. 2016. "Embrace the inner fox: Post-truth as the STS symmetry principle universalized". *Social Epistemology Review & Reply Collective*. https://social-epistemology.com/2016/12/25/embrace-the-inner-fox-post-truth-as-the-sts-symmetry-principle-universalized-steve-fuller/, accessed in October 2019.

Funtowicz, Silvio and Jerome Ravetz. 1993. "Science for the Post-Normal Age". *Futures*, Vol. 25, No. 7, pp. 739–755.

Galbraith, John S. 1975. *Crown and Charter: The Early Years of the British South Africa Company*. Berkeley, University of California Press.

Garforth, Lisa and Tereza Stöckelová. 2012. "Science Policy and STS from Other Epistemic Places". *Science, Technology & Human Values*, Vol. 37, No. 2, pp. 226–240.

Gentle, Leonard. 2008. "Escom to Eskom: From racial Keynesian capitalism to neo-liberalism (1910–1994)". In David A. McDonald (ed). 2008. *Electric Capitalism. Recolonizing Africa on the Power Grid*. Cape Town: HSRC Press, pp. 50–72.

Gibbons, Michael, Limoges, Camille, Nowotny, Helga, Schwartzman, Simon, Scott, Peter and Martin Trow. 1994. *The New Production of Knowledge: The Dynamics of Science and Research in Contemporary Societies*. London: Sage.

Gilroy, Paul. 1993. *The Black Atlantic: Modernity and Double Consciousness*. London: Verso.

Gilson, Norbert. 1994. *Konzepte von Elektrizitätsversonrgung und Elektrizitätswirtschaft: die Entstehung eines neuen Fachgebietes der Technikwissenschaften zwischen 1880 und 1945*. Stuttgart: Verlag für Geschichte der Naturwissenschaften und der Technik.

Glaser, Gustave. 1881. "The Paris Electrical Exhibition". *Science*, Vol. 2, No. 63 (Sep. 10, 1881), pp. 430–433.

Graham, Wayne. 1996. "The Randlord's Bubble 1894–6: South African Gold Mines and Stock Market Manipulation". Discussion Papers in Economic and Social History, No. 10 (August 1996), University of Oxford.

Grandin, Karl, Wormbs, Nina and Sven Widmalm (eds). 2004. *The Science-Industry Nexus: History, Policy, Implications*. Nobel Symposium. Science History Publications/USA and the Nobel Foundation. Sagamore Beach, MA: Watson Publishing International.

Guagnini, Anna. 2014. "A Bold Leap into Electric Light: The Creation of the Società Italiana Edison, 1880–1886". *History of Technology*, Vol. 32, pp. 155–190.

Gugerli, David. 1996. *Redeströme. Zur Elektrifizierung der Schweiz: 1880–1914*. Zürich: Chronos Verlag.

Guggenheim, Michael and Helga Nowotny. 2003. "Joy in Repetition Makes the Future Disappear. A Critical Assessment of the Present State of STS". In Joerges, B. and Nowotny, H. (eds). 2003. *Social Studies of Science and Technology: Looking Back, Ahead*. Dordrecht: Kluwer Academic Publishers, pp. 229–258.

Hackett, Edward J., Amsterdamska, Olga, Lynch, Michael and Judy Wajcman (eds). 2008. *The Handbook of Science and Technology Studies*. Cambridge, MA and London: MIT Press.

Hacking, Ian. 1991. Review: Leviathan and the Air Pump: Hobbes, Boyle and the Experimental Life by Steven Shapin, Simon Schaffer. *The British Journal for the History of Science*, Vol. 24, No. 2, Darwin and Geology (Jun., 1991), pp. 235–241.

Hadley, Arthur E. 1913. "Power Supply on the Rand". *Journal of the Institution of Electrical Engineers*, Vol. 51, No. 220, pp. 2–31.

Hall, Stewart and Bram Gieben (eds). 1992. *Formations of Modernity*. Cambridge: The Open University Press.

Hall, Stewart. 1992. "The West and the Rest. Discourse and Power". In Hall, S. and Gieben, B. (eds). 1992. *Formations of Modernity*. Cambridge: The Open University Press, pp. 275–320.

Hammond, John Hays. 1897. "South Africa and Its Future". *North American Review*, 164 (Jan.–June), 1897.

Hankins, Thomas L. 1986. A Debate over Experiment Review of Leviathan and the Air-Pump by Steven Shapin, Simon Schaffer. *Science, New Series*, Vol. 232, No. 4753 (May 23, 1986), pp. 1040–1042.

Hannah, Leslie. 1984. "On Electrified Societies". Reviewed Work: Thomas P. Hughes, Networks of Power: Electrification in Western Society, 1880–1930. *Isis*, Vol. 75, No. 2 (June, 1984), pp. 379–382.

Hannaway, Owen. 1988. Review: Leviathan and the Air-Pump: Hobbes, Boyle and the Experimental Life by Steven Shapin, Simon Schaffer. *Technology and Culture*, Vol. 29, No. 2 (Apr., 1988), pp. 291–293.

Harding, Sandra. 1998. *Is Science Multicultural?* Bloomington: Indiana University Press.

Harding, Sandra. 2008. *Sciences from Below: Feminisms Postcolonialities, and Modernities*. Durham, NC: Duke University Press.

Harding, Sandra. 2011. *The Postcolonial Science and Technology Studies Reader*. Durham, NC: Duke University Press.

Harvey, Charles. 1989. "The City and International Mining 1879–1914". *Business History Review*, Vol. 63, No. 3 (Autumn 1989), pp. 98–119.

Hatch, Frederick Henry, and John Alexander Chalmers. 1895 [2013]. *Gold Mines of the Rand. Being a Description of the Mining Industry of Witwatersrand, South African Republic.* New York, Cambridge University Press.

Hausman, William J., Hertner, P. and Mira Wilkins. 2008. *Global Electrification. Multinational Enterprise and International Finance in the History of Light and Power 1878–2007.* New York: Cambridge University Press.

Helfferich, Karl. 1921. *Georg von Siemens: Ein Lebensbild aus Deutschlands großer Zeit.* Band 1. Berlin, Julius Springer.

Hess, David J. 1997. *Science Studies. An Advanced Introduction.* New York: New York University Press.

Hess, David J. 2001. "Scientific Culture". In Smelser, N. J. and Baltes, P. B. (eds). 2001. *International Encyclopedia of the Social and Behavioral Sciences.* Oxford: Elsevier Science, pp. 13724–13727.

Hess, David J. 2007. "Crosscurrents: Social Movements and the Anthropology of Science and Technology". *American Anthropologist*, Vol. 109, No. 3, pp. 463–472.

Hill, Christopher. 1986. Review: 'A New Kind of Clergy': Ideology and the Experimental Method. Review of Leviathan and the Air-Pump: Hobbes, Boyle and the Experimental Life by Steven Shapin, Simon Schaffer. *Social Studies of Science*, Vol. 16, No. 4 (Nov., 1986), pp. 726–735.

Hofmänner, Alexandra. 2016. "Science Studies Elsewhere. The Experimental Life and the Other Within". *Social Epistemology: A Journal of Knowledge, Culture and Policy*, Vol. 30, No. 2, pp. 186–212.

Holli, Melvin G. and Peter d'Alroy Jones (eds). 1995 [1977]. *Ethnic Chicago. A Multicultural Portrait.* Fourth edition. Grand Rapids, Michigan: William. B. Eerdmans Publishing Company.

Hughes, Thomas P. 1983. *Networks of Power. Electrification in Western Society, 1889–1930.* Baltimore and London: Johns Hopkins University Press.

Hughes, Thomas P. 1986. "The Seamless Web: Technology, Science, Etcetera, Etcetera". *Social Studies of Science*, Vol. 16, No. 2 (May, 1986), pp. 281–292.

Ilerbaig, Juan. 1992. "The Two STS Sub-Cultures and the Sociological Revolution". *Science, Technology and Society*, Vol. 90 (June 1992), pp. 1–5.

Industrial Commission of Inquiry. 1897. *The Mining Industry: Evidence and Report of the Industrial Commission of Enquiry.* Johannesburg: Transvaal Chamber of Mines.

Inwood, Stephen. 2005. *City of Cities. The Birth of Modern London.* London: Macmillan.

Irving, Sarah. 2008. Natural Science and the Origins of the British Empire. London: Pickering and Chatto.

Jacob, Margaret C. 1986. Review: Leviathan and the Air Pump: Hobbes, Boyle, and the Experimental Life by Steven Shapin, Simon Schaffer. *Isis*, Vol. 77, No. 4 (Dec., 1986), pp. 719–720.

Jacobs, A. M. 1941. "The Development of the Electrical Power Supply in the Union of South Africa". *The Transactions of the South African Institute of Electrical Engineers*, August 1941.

Jamison, Andrew. 1987. "National Styles of Science and Technology: A Comparative Model". *Sociological Inquiry*, Vol. 57, No. 2 (Spring), pp. 144–58.

Jasanoff, Sheila and Hilton R. Simmet. 2017. "No funeral bells: Public reason in a 'post-truth' age". *Social Studies of Science*, Vol. 47, No. 5, pp. 751–770.

Jennings, Richard C. 1988. Review: Leviathan and the Air-Pump: Hobbes, Boyle, and the Experimental Life by Steven Shapin, Simon Schaffer. *The British Journal for the Philosophy of Science*, Vol. 39, No. 3 (Sep., 1988), pp. 403–410.

Jones, J.D.F. 1995. *Through Fortress and Rock. The Story of Gencor 1895–1995*. Johannesburg, Jonathan Ball Publishers.

Kale, Sunila S. 2014. "Structures of Power. Electrification in Colonial India". *Comparative Studies of South Asia, Africa and the Middle East*, Vol. 34, No. 3, pp. 454–475.

Kargon, Robert. 1986. Review: Leviathan and the Air-Pump: Hobbes, Boyle, and the Experimental Life by Steven Shapin, Simon Schaffer. *Albion: A Quarterly Journal Concerned with British Studies*, Vol. 18, No. 4 (Winter, 1986), pp. 665–666.

Katz, Elaine N. 1999. "Revisiting the Origins of the Industrial Colour Bar in the Witwatersrand Gold Mining Industry, 1891–1899". *Journal of Southern African Studies*, Vol. 25, No. 1, pp. 73–97.

Klawitter, Nils. 2013. "Die Geldmaschine". *Spiegel Geschichte*, 3, 2013 (28 May), pp. 78–81.

Klingenberg, Georg. 1913. "Electricity Supply in Large Cities". *The Electrician*, 72, pp. 398–400.

Klingenberg, Georg. 1916 [1913]. *Large Electric Power Stations, their Design and Construction, with Examples of Existing Stations*. New York: D. Van Nostrand Co.

Knauft, Bruce M. (ed). 2002. *Critically Modern? Alternatives, Alterities, Anthropologies*. Bloomington: Indiana University Press.

Knorr-Cetina, Karin. 1999. *Epistemic Cultures: How the Sciences Make Sense*. Chicago: Chicago University Press.

Kubicek, Robert V. 1979. *Economic Imperialism in Theory and Practice. The Case of South African Gold Mining Finance, 1886–1914*. Durham (N.C.): Duke University Press.

Kuhn, Thomas S. 1962 [1970]. *The Structure of Scientific Revolutions*. Chicago: University of Chicago Press.

Latour, Bruno. 1993. *We Have Never Been Modern*. Cambridge, MA.: Harvard University Press.

Latour, Bruno. 1999. "For David Bloor ... and Beyond: A Reply to David Bloor's 'Anti-Latour". *Studies in the History and Philosophy of Science*, Vol. 30, No. 1, pp. 113–129.

Latour, Bruno. 2007. "The Recall of Modernity: Anthropological Approaches". *Cultural Studies Review*, Vol. 13, No. 1, pp. 11–30.

Ley, Barbara L. and Paul R. Brewer. 2018. "Social Media, Networked Protest, and the March for Science". *Social Media and Society*, Vol. 4, No. 3., pp. 1–12.

Longino, Helen. 2001. *The Fate of Knowledge*. Princeton: Princeton University Press.

Lynas, Mark. 2017. Scientists to take to the streets in global march for truth. *The Guardian*, 17 April 2017.

Lynch, Michael (ed). 2012. *Science and Technology Studies, Critical Concepts in the Social Sciences*. 4 Volumes. London and New York: Routledge.

Mackenzie, David and Judy Wajcman (eds). 1999 [1985]. *The Social Shaping of Technology*. Buckingham and Philadelphia: Open University Press.

Maguire, James Rochfort. 1896: *The Pioneers of Empire*. London: Methuen & Co.

Malcolm, Noel. 1981. "Hobbes, Sandys, and the Virginia Company". *The Historical Journal*, Vol. 24, No.2 (1981), pp. 297–321.

Mamdani, Mahmood. 1996. *Citizen and subject. Contemporary Africa and the Legacy of Late Colonialism*. Princeton: Princeton University Press.

Marquard, Andrew. 2006. *The Origins and Development of South African Energy Policy*. PhD Thesis, Faculty of Engineering and the Built Environment, University of Cape Town.

Martin, Ben R., Paul Nightingale and Alfredo Yegros-Yegros. 2012. "Science and Technology Studies: Exploring the Knowledge Base". *Research Policy*, Vol. 41, No. 7, (September 2012), pp. 1182–1204.

McDonald, John F. 2016. *Chicago: An Economic History*. Routledge Advances in Regional Economics, Science and Policy. New York, Routledge.

McGregor, JoAnn. 2003. "The Victoria Falls 1900–1940: Landscape, Tourism and the Geographical Imagination". *Journal of Southern African Studies*, Vol. 29, No. 3 (Sep., 2003), pp. 717–737.

McNeil, Maureen. 2005. "Introduction: Postcolonial Technoscience". *Science as Culture*, Vol. 14, No. 2, pp. 105–112.

Mels, Edgar. 1900. "The Natives of South Africa". *Scientific American. A Weekly Journal of Practical Information, Art, Science, Mechanics, Chemistry, and Manufactures*. Vol. LXXXII, No. 4 (January 27), pp. 56–58.

Merrington, Peter. 2001. "A Staggered Orientalism: The Cape-to-Cairo Imaginary". *Poetics Today*, Vol. 22, No. 2 (Summer 2001), pp. 323–364.

Merton, Robert K. 1973. *The Sociology of Science: Theoretical and Empirical Investigations*. Chicago: University of Chicago Press.

Merz, Charles H. and William McLellan. 1919. *South African Railways: Report on the Introduction of Electric Traction.*

Merz, Charles H. and William McLellan. 1920. *Electric Power Supply in the Union of South Africa.* Report to the Prime Minister of the Union of South Africa. April, 1920.

Miles, Kate. 2013. *The Origins of International Investment Law: Empire, Environment and the Safeguarding of Capital.* Cambridge: Cambridge University Press.

Misa, Thomas J. 2004. *Leonardo to the Internet. Technology and Culture from the Renaissance to the Presence.* Baltimore: Johns Hopkins University Press.

Misa, Thomas J., Brey, Philip and Andrew Feenberg (eds). 2003. *Modernity and Technology.* Cambridge MA.: MIT Press.

Mitchell, Timothy (ed). 2000. *Questions of Modernity.* Minneapolis: University of Minnesota Press.

Mitchell, Timothy. 2000. "The Stage of Modernity". In Mitchell, T. (ed). 2000. *Questions of Modernity.* Minneapolis: University of Minnesota Press, pp. 1–34.

Mitchell, T. 2000 [1992]. "Orientalism and the Exhibitionary Order". In Dirks, Nicholas B. (ed). 2000. *Colonialism and Culture.* Ann Arbor: University of Michigan Press, pp. 289–318.

Moloney, Pat. 2011. "Hobbes, Savagery, and International Anarchy". *American Political Science Review*, Vol. 105, No. 1, pp. 189–204.

Morton, David L. 2002. "Reviewing the history of electric power and electrification". *Endeavour*, Vol. 26, No. 2 (1 June 2002), pp. 60–63.

Mudimbe, Valentin, Y. 1988. *The Invention of Africa: Gnosis, Philosophy, and the Order of Knowledge.* London: James Currey.

Nakajima, Hideto. 2007. "Differences in East Asian STS: European Origin or American Origin?" *East Asian Science, Technology and Society: An International Journal*, Vol. 1, No. 2, pp. 237–241.

National Research Council. 2011. *U.S. and International Perspectives on Global Science Policy and Science Diplomacy.* Report of a Workshop. Committee on Global Science Policy and Science Diplomacy. NAS (National Academy of Sciences) Press.

Native Affairs Commission. 1905. *South African Native Affairs Commission, 1903–1905. Report with Annexures Nos 1–9.* Cape Town: Cape Times Limited.

Newbury, Colin. 2009. "Cecil Rhodes, De Beers and Mining Finance in South Africa: The Business of Entrepreneurship and Imperialism". In Dumett, Raymond E. (ed). 2009. *Mining Tycoons in the Age of Empire, 1870–1945. Entrepreneurship, High Finance, Politics and Territorial Expansion.* Farnham: Ashgate, pp. 85–108.

Nkosi, N. K. 1987. "American Mining Engineers and the Labor Structure in the South African Gold Mines". *African Journal of Political Economy*, Vol. 1, No. 2, pp. 63–81.

North, J. D. 1987. Review: Leviathan and the Air-Pump: Hobbes, Boyle, and the Experimental Life by Steven Shapin, Simon Schaffer. *American Scientist*, Vol. 75, No. 2 (March–April 1987), p. 216.

Nowotny, Helga, Scott, Peter and Michael Gibbons. 2001. *Re-Thinking Science. Knowledge and the Public in an Age of Uncertainty*. Cambridge: Polity Press.

Nowotny, Helga. 2007. "How many policy rooms are there? Evidence-based and other kinds of science policies". *Science, Technology and Human Values*, Vol. 32, No. 4, pp. 479–490.

Nye, David E. 1992. *Electrifying America: Social Meanings of a New Technology*. Cambridge, MA: MIT Press.

Nye, Joseph S. 1990. "Soft power". *Foreign Policy*, No. 80, pp. 153–171.

OECD. 2005. *Forum on the Internationalisation of R&D. Background Paper: Internationalisation of R&D-trends, issues and implication for S&T policies*. Paris: OECD Publishing.

OECD. 2010a. *Measuring Innovation: A New Perspective*. Paris: OECD Publishing.

OECD. 2010b. *Report on Establishing Large International Research Infrastructures: Issues and Opinions*. Paris: OECD Publishing.

OECD. 2011. *OECD Science, Technology and Industry Scoreboard 2011: Innovation and Growth in Knowledge Economies*. Paris: OECD Publishing.

OECD. 2012. *Meeting Global Challenges through Better Governance: International Cooperation in Science, Technology and Innovation*. Paris: OECD Publishing.

Oliver, Roland and G. N. Sanderson (eds). 1985. *The Cambridge History of Africa*, Volume 6. Cambridge: Cambridge University Press.

Penner C. D. 1940. "Germany and the Transvaal before 1896". *The Journal of Modern History*, Vol. 12, No.1, pp. 31–58.

Phillips, Lionel. 1905. *Transvaal Problems: Some Notes on Current Politics*. London: J. Murray.

Pickering, Andrew. 1994. Review: We Have Never Been Modern. *Modernism/Modernity*, Vol. 1, No. 3. (September 1994), pp. 257–8.

Pinner, Felix. 1918. *Emil Rathenau und das Elektrische Zeitalter*. Leipzig: Akademische Verlagsgesellschaft.

Polanyi, Michael. 1958. *Personal Knowledge: Towards a Post-Critical Philosophy*. Chicago: University of Chicago Press.

Prasad, Amit. 2008. "Science in Motion: What Postcolonial Science Studies Can Offer". *Electronic Journal of Communication Information & Innovation in Health*, Vol. 2, No. 2, pp. 35–47.

Prasad, Amit. 2014: *Imperial Technoscience. Transnational Histories of MRI in the US, Britain, and India*. London: MIT Press.

Pratt, Mary Louise. 1992. *Imperial Eyes. Studies in Travel Writing and Transculturation*. London: Routledge.

Price, Bernard. 1915. "The Power Supply to the Mines of the Rand". *SA Mining Journal*, 27 Feb. 1915, pp. vi–xii.

Price, Bernard. 1916. "The Power Supply of the Rand. Inaugural Address". *Journal of the South African Institution of Engineers*, Aug. 1916, pp. 61–63.

Price, D.J. de Solla. 1963. *Little Science, Big Science*. New York: Columbia University Press.

Public Relations Office. 1967. "City of Johannesburg. Brief History of the Development of its System of Government". Public Relations Office, City Hall, Johannesburg.

Randeria, Shalini. 1999a. "Geteilte Geschichte und verwobene Moderne". In Rüsen, J., Leitge, H., Jegelka, N., Aïtmatov C. (eds). 1999. *Zukunftsentwürfe. Ideen für eine Kultur der Veränderung*. Frankfurt a.M.: Campus, pp. 87–96.

Randeria, Shalini. 1999b. "Jenseits von Soziologie und soziokultureller Anthropologie. Zur Ortsbestimmung der nichtwestlichen Welt in einer zukünftigen Sozialtheorie". *Soziale Welt*, Vol. 50, No. 4, pp. 373–382.

Randeria, Shalini. 2002. "Entangled Histories of Uneven Modernities. Civil Society, Caste Solidarities and the Post-Colonial State in India". In Elkana, Y., Krastev, I., Macamo, E., and S. Randeria (eds). 2002. *Unravelling Ties: From Social Cohesion to New Practices of Connectedness*. Frankfurt a.M./ New York: Campus/St. Martin's Press, pp. 284–311.

Rathenau, Walther. 1908. *Reflexionen*. Leipzig: S. Hirzel.

Redfield, Peter. 2002. "The Half-Life of Empire in Outer Space". *Social Studies of Science*, Vol. 32, No. 5/6, pp. 791–825.

Restivo, Sal. 1987. "Science Studies – What Is To Be Done". *Science, Technology, & Human Values*, Vol. 12, No. 2 (Spring, 1987), pp. 13–18.

Restrepo, Eduardo and Arturo Escobar. 2005. "Other Anthropologies and Anthropology Otherwise. Steps to a World Anthropologies Framework". *Critique of Anthropology*, Vol. 25, No. 2, pp. 99–129.

Richard, Thomas A. 1922. *Interviews with Mining Engineers*. San Francisco: Mining & Scientific Press.

Richardson, Peter and Jean Jacques Van Helten. 1982. "Labour in the South African Gold Mining Industry, 1886–1914". In Marks, Shula and Richard Rathbone (eds). 1982. *Industrialisation and Social Change in South Africa – African Class Formation, Culture, and Consciousness, 1870–1930*. New York: Longman, pp. 77–98.

Rider, J.H. 1915. "The Power Supply of the Central Mining-Rand Mines Group". *The Journal of the Institution of Electrical Engineers*, Vol. 53, No. 247 (1 May), pp. 609–632.

Rip, Arie. 1994. "The Republic of Science in the 1990s". *Higher Education*, Vol. 28, pp. 3–23.

Rip, Arie. 1999. "STS in Europe". *Science, Technology and Society*, Vol. 4, No. 1, pp. 73–80.

Rodriguez, Encarnacion, G., Boatcâ, Manuela and Sérgio Costa (eds). 2010. *Decolonizing European Sociology: Transdisciplinary Approaches.* Surrey: Ashgate Publishing.

Rosenbach, Harald. 1993. *Das Deutsche Reich, Großbritannien und der Transvaal (1896–1902); Anfänge deutsch-britischer Entfremdung.* Band 52. Göttingen: Vandenhoeck & Ruprecht.

Ross, Ashley D., Struminger, Rhonda, Winking, Jeffrey and Kathryn R. Wedemeyer-Strombel. 2018. "Science as a Public Good: Findings from a Survey of March for Science Participants". *Science Communication,* Vol 40, No. 2, pp. 228–245.

Rowland, John. 1960. *Progress in Power: the Contribution of Charles Merz and His Associates to Sixty Years of Electrical Development, 1899–1959.* London: Newman Neame.

Royal Society. 2010. "New Frontiers in Science Diplomacy: Navigating the Changing Balance of Power". A Royal Society Policy Document, January 2010.

Royal Society. 2011. "Knowledge, Networks and Nations. Global Scientific Collaboration in the 21st Century". The Royal Society, London.

Said, Edward W. 2003 [1977]. *Orientalism. Western Conceptions of the Orient.* London: Penguin.

Sanford, Greg. 1989. "Illuminating Systems: Edison and Electrical Incandescence". *OAH Magazine of History,* Vol. 4, No. 2 (Spring 1989), pp. 16–19.

Secord, James A. 2004. "Knowledge in Transit". *Isis,* Vol. 95, pp. 654–672.

Segreto, Luciano. 1994. "Financing the Electric Industry Worldwide: Strategy and Structure of the Swiss Electric Holding Companies, 1895–1945". *Business and Economic History,* Vol. 23 (Fall 1994), pp. 162–75.

Seth, Suman. 2005. "Putting knowledge in its place: Science, colonialism, and the postcolonial". *Postcolonial Studies,* Vol. 12, No. 4, pp. 373–388.

Seth, Suman. 2017. "Colonial History and Postcolonial Science Studies". *Radical History Review,* No. 127, pp. 63–85.

Shamir, Ronen. 2013. *The Electrification of Palestine.* Stanford CA: Stanford University Press.

Shapin, Steven and Simon Schaffer. 1989 [1985]. *Leviathan and the Air-Pump: Hobbes, Boyle, and the Experimental Life.* Princeton: Princeton University Press.

Shapin, Steven and Simon Schaffer. 2011. "Up for Air: Leviathan and the Air-Pump a Generation On". Introduction to the 2011 Edition. In Shapin S. and S. Schaffer. 2011 [1985]: *Leviathan and the Air-Pump: Hobbes, Boyle, and the Experimental Life.* Princeton: Princeton University Press.

Sismondo, Sergio. 2008. "Science and Technology Studies and an Engaged Program". In Hacket, E. Amsterdamska, O., Lynch, M. and Wajcman, J. (eds). 2008. *The Handbook of Science and Technology Studies.* Cambridge and London: MIT Press, pp. 13–17.

Sismondo, Sergio. 2010 [2004]. *An Introduction to Science and Technology Studies.* West-Sussex: Wiley-Blackwell.

Sismondo, Sergio. 2017. "Post-truth?". *Social Studies of Science,* Vol. 47, No. 1, pp. 3–6.

Skinner, Walter R. 1911. *The Mining Manual 1911.* London: Walter R. Skinner and Financial Times.

Sothen, Hans von. 1915. *Die Wirtschaftspolitik der Allgemeinen Elektrizitäts-Gesellschaft.* PhD Thesis, Freiburg.

Spiegel-Rösing, Ina and Derek de Solla Price (eds). 1977. *Science, Technology and Society: A Cross-Disciplinary Perspective.* London: Sage.

Spivak, Gayatri Chakravorty. 1988. "Can the Subaltern Speak?". In Nelson, C. and L. Grossberg (eds). 1988. *Marxism and the Interpretation of Culture.* Urbana/Chicago: University of Illinois Press, pp. 271–316.

Stein, Josephine A. 2002. "Globalisation, Science, Technology and Policy". *Science and Public Policy,* Vol. 29, No. 6 (December 2002), pp. 402–408.

Stevenson, Angus (ed). 2010. *Oxford Dictionary of English.* Third Edition. Oxford, Oxford University Press.

Sulzberger, Carl. 2014. "The Pearl Street Generating Station, 1882". *IEEE Global History Network,* 2014. Web Article. http://ethw.org/Milestones:Pearl_Street_Station.

Swidler, Anne and Jorge Arditi. 1994. "The New Sociology of Knowledge". *Annual Review of Sociology,* Vol. 20 (1994), pp. 305–329.

Swiss Federal Council. 2010. "Internationale Strategie der Schweiz im Bereich Bildung, Forschung und Innovation (BFI)". SBF, Bern.

The New York Times. 1912. "Power Plant Men Here From Rhodesia". Sunday, October 6, 1912.

Thompson, Charis. 2008. "Medical Tourism, Stem Cells, Genomics: EASTS, Transnational STS, and the Contemporary Life Sciences". *East Asian Science, Technology and Society: an International Journal,* Vol. 2, No. 3, pp. 433–438.

Transvaal Native Affairs Department. 1910. *Annual Report 1909.* Pretoria: Government Printing and Stationary Office.

Traynham, James G. 1987. Review: Leviathan and the Air-Pump: Hobbes, Boyle, and the Experimental Life by Steven Shapin, Simon Schaffer. *The Journal of Interdisciplinary History,* Vol. 18, No. 2 (Autumn, 1987), pp. 351–353

Troost, N. and H.B. Norman. 1969. "Electricity Supply in South Africa 1909–1969". *The Transactions of the S.A. Institute of Electrical Engineers,* September 1969, pp. 177–192.

Trouillot, Michel-Rolph. 1991. "Anthropology and the Savage Slot: The Poetics and Politics of Otherness". In Fox, R. G. (ed). 1991. *Recapturing Anthropology: Working in the Present.* Santa Fe, N.M.: School of American Research Press, pp. 17–44.

Trouillot, Michel-Rolph. 2002a. "The Otherwise Modern. Caribbean Lessons from the Savage Slot". In Knauft, B. M. (ed). 2002. *Critically Modern? Alternatives, Alterities, Anthropologies*. Bloomington: Indiana University Press, pp. 220–237.

Trouillot, Michel-Rolph. 2002b. "North Atlantic Universals: Analytical Fictions, 1492–1945". *The South Atlantic Quarterly*, Vol. 101, No. 4, pp. 839–858.

Trouillot, Michel-Rolph. 2003. *Global Transformations: Anthropology and the Modern World*. New York: Palgrave Macmillan.

Tshitereke, Clarence. 2006. *The Experience of Economic Redistribution: The Growth, Employment and Redistribution Strategy in South Africa*. New York: Routledge.

Tuffnell, Stephen. 2015. "Engineering Inter-Imperialism: American miners and the transformation of global mining: 1871–1910". *Journal of Global History*, Vol. 10, pp. 53–76.

Turner, Stephen. 2008. "The Social Study of Science Before Kuhn". In Hackett, E. J., Amsterdamska, O., Lynch, M., and J. Wajcman (eds). 2008. *The Handbook of Science and Technology Studies*. Cambridge, MA and London: MIT Press, pp. 33–62.

Turrell, Robert Vicat and Jean-Jacques Van Helten. 1986. "The Rothschilds, the Exploration Company and Mining Finance". *Business History*, Vol. 28, No. 2 (April 1986), pp. 183–186.

UNESCO. 2010. *Unesco Science Report 2010: The Current Status of Science around the World*. Paris: Unesco Publishing.

Van Alphen, Ernst. 1991. "The Other Within". In Corbey, R. and Leerssen, J. (eds). 1991. *Alterity, Identity, Image. Selves and Others in Society and Scholarship*. Amsterdam: Rodopi, pp. 1–16.

Van der Poel, Jean (ed). 1977. *Selections from the Smuts papers*. Vol. VIII (August 1945–October 1950). Cambridge, Cambridge University Press.

Van Helten, Jean Jacques. 1982. "Empire and high finance: South Africa and the international gold standard 1890–1914". *The Journal of African History*, Vol. 23, No. 4 (October 1982), pp 529–548.

Veeser, Cyrus. R. 2013. "A Forgotten Instrument of Global Capitalism? International Concessions, 1870–1930". *International History Review*, Vol. 35, No. 5, pp. 1136–1155.

Verschoyle, F. (ed). 1900. *Cecil Rhodes: His Political Life and Speeches, 1881–1900*. London: Chapman and Hall.

Voestermans, Paul. 1991. "Alterity/Identity: A Deficient Image of Culture". In Corbey, R., and Leerssen, J. (eds). 1991. *Alterity, Identity, Image. Selves and Others in Society and Scholarship*. Amsterdam: Rodopi, pp. 219–250.

Waks, Leonard J. 1993. "STS as an Academic Field and a Social Movement". *Technology in Society*, Vol. 15, No. 4, 1993, pp. 399–408.

Webster, Andrew. 1991. *Science, Technology, Society. New Directions*. New Brunswick NJ: Rutgers University Press.

Webster, Andrew. 2007. "Crossing Boundaries: Social Science in the Policy Room". *Science Technology Human Values*, Vol. 32, No. 4, pp. 458–478.

Weinberger, Gerda. 1971. "Das Victoria-Falls-Power-Projekt der AEG und die Deutsche Kapitaloffensive in Südafrika vor dem ersten Weltkrieg. Zur Rolle von Industrie- und Bankkaptial in der ökonomischen Expansion und der Entwicklungs staatsmonopolistischer Formen". *Jahrbuch für Wirtschaftsgeschichte*, pp. 57–83.

Weinberger, Gerda. 1975. "An den Quellen der Apartheid. Studien über koloniale Ausbeutungs- und Herrschaftsmethoden in Südafrika und die Zusammenarbeit des Deutschen Imperialismus mit dem Englischen Imperialismus und den Burischen Nationalisten (1902–1914)". Forschungen zur Wirtschaftsgeschichte. Berlin: Akademie-Verlag.

Westfall, Richard S. 1987. Review: Leviathan and the Air-Pump: Hobbes, Boyle, and the Experimental Life by Steven Shapin, Simon Schaffer. *Philosophy of Science*, Vol. 54, No. 1 (Mar., 1987), pp. 128–130.

Wheatcroft, Geoffrey. 1985. *The Randlords: The Men Who Made South Africa*. London: Weidenfeld & Nicolson.

Wilkins, Mira. 1988. "The Free-Standing Company, 1870–1914: An Important Type of British Foreign Direct Investment". *The Economic History Review*, Vol. 41, No. 2 (May, 1988), pp. 259–282.

Wilkins, Mira. 1998. "The Free-Standing Company Revisited". In Wilkins, M. and Schröter, H. (eds). 1998. *The Free-Standing Company in the World Economy 1839–1996*. New York: Oxford University Press, pp. 3–64.

Wills, Walter H. and R. J. Barrett (eds). 1905. *The Anglo-African Who's Who and Biographical Sketch-book*. London: George Routledge & Sons.

Wills, Walter H. (ed). 2006 [1907]. *The Anglo-African Who's Who and Biographical Sketch-book 1907*. London: Jeppestown Press.

Wilson, Francis. 1972. *Labour in the South African Gold Mines 1911–1969*. Oxford: Cambridge University Press.

Winner, Langdon. 1993. "Upon Opening the Black Box and Finding it Empty. Social Constructivism and the Philosophy of Technology". *Science, Technology and Human Values*, Vol. 18, No. 3 (Summer 1993), pp. 362–378.

Woolgar, Steve. 1991. "The Turn to Technology in Social Studies of Science". *Science Technology Human Values*, Vol. 16, No. 1 (January 1991), pp. 20–50.

Yearley, Steven. 2005. *Making Sense of Science: Understanding the Social Study of Science*. London: Sage.

Schwabe Verlag's signet was
Johannes Petri's printer's mark.
His printing workshop was
established in Basel in 1488 and
was the origin of today's Schwabe
Verlag. The signet refers back to
the beginnings of the printing press,
and originated in the entourage of
Hans Holbein. It illustrates a verse of
Jeremiah 23:29: 'Is not my word
like fire, says the Lord, and like a
hammer that breaks a rock in pieces?'